EPISODES
FROM
THE EARLY HISTORY
OF MATHEMATICS

NEW MATHEMATICAL LIBRARY

published by

Random House and The L. W. Singer Company

for the

Monograph Project

of the

SCHOOL MATHEMATICS STUDY GROUP†

EDITORIAL PANEL

† The School Mathematics Study Group represents all parts of the mathematical profession and all parts of the country. Its activities are aimed at the improvement of teaching of mathematics in our schools. Further information can be obtained from: School Mathematics Study Group, Cedar Hall, Stanford University, Stanford, California.

EPISODES

FROM

THE EARLY HISTORY
OF MATHEMATICS

by

Asger Aaboe

Yale University

13

RANDOM HOUSE
THE L. W. SINGER COMPANY

Acknowledgments

Figures 1.1, 1.3, 1.5 are reproduced from *The Babylonian Expedition of the University of Pennsylvania*, Series A: Cuneiform Texts, Vol. XX, Part 1, by H. V. Hilprecht, Philadelphia, 1906, through the courtesy of the Department of Archeology, University of Pennsylvania.

Figure 1.4 is reproduced from A. Parrot's *Le Palais de Mari, Architecture*, Librairie Orientaliste Paul Geuthner, Paris, 1958, through the courtesy of Mission Archeologique de Mari.

Figure 1.6a is reproduced with the kind permission of Professor Hallo of the Yale Babylonian Collection.

Illustrated by the author

First Printing

To

Kirsten, Anne,

Erik, and Niels.

NEW MATHEMATICAL LIBRARY

Other titles will be announced as ready

Note to the Reader

This book is one of a series written by professional mathematicians in order to make some important mathematical ideas interesting and understandable to a large audience of high school students and laymen. Most of the volumes in the *New Mathematical Library* cover topics not usually included in the high school curriculum; they vary in difficulty, and, even within a single book, some parts require a greater degree of concentration than others. Thus, while the reader needs little technical knowledge to understand most of these books, he will have to make an intellectual effort.

If the reader has so far encountered mathematics only in classroom work, he should keep in mind that a book on mathematics cannot be read quickly. Nor must he expect to understand all parts of the book on first reading. He should feel free to skip complicated parts and return to them later; often an argument will be clarified by a subsequent remark. On the other hand, sections containing thoroughly familiar material may be read very quickly.

The best way to learn mathematics is to *do* mathematics, and each book includes problems, some of which may require considerable thought. The reader is urged to acquire the habit of reading with paper and pencil in hand; in this way mathematics will become increasingly meaningful to him.

For the authors and editors this is a new venture. They wish to acknowledge the generous help given them by the many high school teachers and students who assisted in the preparation of these monographs. The editors are interested in reactions to the books in this series and hope that readers will write to: Editorial Committee of the NML series, in care of THE COURANT INSTITUTE OF MATHEMATICAL SCIENCES, NEW YORK UNIVERSITY, New York 3, N. Y.

The Editors

Contents

Introduction

If a schoolboy suddenly finds himself transplanted to a new school in foreign parts, he is naturally puzzled by much of the curriculum. The study of languages and of subjects strongly depending on language, such as literature, changes radically from nation to nation, and some subjects, history for one, may even be interpreted differently in different parts of a single country. But in the sciences and in mathematics the boy will probably be quite at home; for, even though order and fashion of presenting details may vary from place to place, these subjects are essentially international.

But if we now imagine our schoolboy transported not only to a different place but also to a different age—say to Greece two thousand years ago, or Babylonia four thousand years ago—he would have to look hard to find anything that he could recognize as science, either in content or in method. What was called "physics" in Aristotle's day, with its discussions of the number of basic principles and of the nature of motion, we would classify as philosophy; and its connection with modern physics appears only after a careful study of the development of the physical sciences. Mathematics alone would now look familiar to our schoolboy: he could solve quadratic equations with his Babylonian fellows and perform geometrical constructions with the Greeks. This is not to say that he would see no differences, but they would be in form only, and not in content; the Babylonian number system was not the same as ours, but the Babylonian formula for solving quadratic equations is still in use.

The unique permanence and universality of mathematics, its independence of time and cultural setting, are direct consequences of its very nature. In Chapter 2 I shall say something about the structure of mathematical theories, so here I will content myself with drawing attention to only a few facets of our subject's singular character.

1

First I must mention that mathematics is cumulative; that is, it never loses territory, and its boundaries are ever moving outwards. This is in part a consequence of its absolute standards which ensure that what once is good mathematics will always be so and will remain part of the living body of mathematical knowledge. This steady growth offers a contrast to the progress of physics, to take but one example, which has been the victim, or rather beneficiary, of several radical revolutions. So, while Greek physics has only historical interest for a modern physicist, Greek mathematics is still good mathematics which is unavoidable for a modern mathematician. It was the English mathematician Littlewood who said, with a donnish simile, that we should think of the Greek mathematicians not as clever schoolboys or "scholarship candidates," but as "Fellows of another college."

Another facet I must mention is the deductive character of mathematics: a mathematical theory progresses in an orderly, logical fashion from explicitly stated axioms. One consequence of this is that knowledge of a certain theorem implies, or should imply, knowledge of all the theorem's predecessors linking it to the axioms. A beginner must then begin at the beginning, and the beginning is often old in substance. I can illustrate this point with a biological dictum which, because of its curious phrasing, has stuck in my mind. It says that *ontogeny recapitulates phylogeny*, and it means that in the development of an individual we see, in swift review, the development of its entire species. If taken literally this dictum can lead, and has led, to all sorts of nonsense, but properly qualified it contains a truth. In the same modified sense it applies to the species of mathematicians. The embryonic development of a mathematician, that is, the education which leads him from the beginnings up to the research front of his day, indeed follows crudely the development of mathematics itself.

Thus, whether we want it or not, the past is very much with us in mathematics, and, whether he wants to or not, a mathematician must begin by studying what in substance is ancient mathematics, in whatever garb the mathematical fashion may dictate. Also, mathematicians are justly proud of the high antiquity of their subject: mathematics is so ancient a discipline that even the study of its history became a recognized field of scholarly endeavour long before most of the sciences. It is therefore particularly natural for students of mathematics to acquaint themselves with the history of their subject, and it is the purpose of the present little volume to help them do so.

I have chosen not to attempt a survey of the history of mathematics from its beginnings to the present. Such a treatment, when confined to a reasonable length, is necessarily weak in mathematical detail and is meaningful only to those who are proficient enough in mathematics to supply depth to a shallow picture. Instead I have selected four episodes from the early history of mathematics and treated them in detail, with comments to convey some notion of their proper setting. As guiding principles for my choice of topics I have used first, that their mathematical content should be within reach of a student with knowledge of high school algebra and geometry. So I have excluded anything that has to do with limit processes and calculus (except the short and elegant argument that led Archimedes to his discovery of the volume and surface area of a sphere, and which I could not resist including). Further, I wanted my selections to be mathematically significant, representative of their periods and authors, and yet off the track beaten by popular histories of mathematics; I wanted them capable of independent treatment, yet having some themes and ideas in common.

Of course, such goals can only be approximated. The topics I came up with are, in order of their appearance in this book and also in chronological order, a presentation of Babylonian mathematics recovered from cuneiform texts only during the last half century; Euclid's construction of the regular pentagon from his *Elements*; three small samples of Archimedes' mathematics: his trisection of an angle, his construction of the regular heptagon, and his discovery of the volume and surface of a sphere; and, lastly, Greek trigonometry as it is presented by Ptolemy in his *Almagest*. I have endeavoured throughout to emphasize what the sources of our knowledge of ancient mathematics are, and in my presentation of the material I have tried to stay as close to the texts as is comfortable for a modern reader.

A recurrent theme in the selections from Greek mathematics is the problem of dividing the circle into a number of equal parts; Euclid achieves the division into five parts by compasses and straightedge alone, Archimedes must employ more complicated tools, and Ptolemy is interested in computing the length of the chord subtending a proper part of the circumference of a circle. The Babylonian number system, which was the backbone of Babylonian mathematics, is adopted by Ptolemy as the only reasonable manner of expressing fractions (and is hence preserved in our subdivisions of degrees and hours). Babylonian influence may be detected in Euclid's formulation of quadratic equations, and though his method of solution differs on the surface from that of the Babylonians, there are similarities in the two approaches to the same problem. I shall leave it to the reader to discover other connecting threads between the four chapters, though each can be read separately.

Finally, I wish to make two apologetic and warning remarks about Greek names in the last three chapters. First, I have made no attempt at consistency in their spelling, but have simply written down what came naturally to my pen. If a reader should be interested in the proper Greek form of a particular name, he can readily reconstruct it from my spelling; consistency would prohibit such time-honoured usages as *Plato*, *Aristotle*, and *Euclid*. Second, the number of names of Greek mathematicians and scholiasts who make but one or two insignificant appearances each in my tale is large, and it might well be argued that they were better omitted. But whenever I had the choice of writing, for example, "Stobaeus tells" or "we have it on ancient authority," I chose the former alternative, for I see no excuse for imprecision when precision is so easily attained. The reader who wishes to look up the reference is helped by my choice, and he who does not is not harmed. I have, however, not wanted to clutter up the pages of this book with any more detailed learned apparatus; at any rate, several of the works in the bibliography at the end contain exhaustive references.

There is great excitement in discovering the patterns of thought of great minds of the distant past, and in the mathematical sciences one can recognize when resonance is achieved with a much higher degree of certainty than anywhere else. It is a privilege to show others along paths first trodden so long ago, or, in a fine old phrase, to make the lips of the ancients move in their graves. There is, however, no real substitute for reading the old mathematicians themselves, and if this little book should induce some of its readers to do so, it will have served its purpose well.

CHAPTER ONE

Babylonian Mathematics

1.1 Sources

When we speak of Babylonian mathematics we mean the kind of mathematics cultivated in ancient Mesopotamia—the country between the rivers Euphrates and Tigris or, roughly, what is known as Iraq today. We are therefore using the term Babylonian in a wider sense than is customary in accounts of the political history of the Near East, where it refers to the state about the city Babylon.

Until quite recently one knew of Babylonian mathematics only through scattered references in the classical Greek literature to Chaldean, i.e. Babylonian, mathematicians and astronomers. On the basis of these references it was assumed that the Babylonians had had some sort of number mysticism or numerology; but we now know how far short of the truth this assumption was.

In the latter part of the nineteenth century archeologists began digging in the ancient city mounds in Mesopotamia. These mounds are made up of the debris of the long-lived cities of the past. The houses were built mostly of unbaked brick (as they often are even today), and every rainfall washed a bit of them off. New houses were built on the same sites and little by little the ground level rose until the present mounds were formed. This process is still going on, for some of these city mounds are even now crowned by inhabited villages, direct descendants of ancient cities. Thus, if we make a vertical cross-section of a mound, we find layer upon layer of different stages of the same city, the oldest at the bottom.

Excavations of the mounds yielded, among much other evidence of the splendid ancient civilizations, thousands upon thousands of clay tablets with writing upon them. It was recognized early that some of these had to do with numbers, but it was not until some thirty years ago that a thorough understanding and appreciation of Babylonian mathematics was achieved.

We have now some 400 tablets and fragments of tablets of mathematical content which have been carefully copied, transcribed, translated, and explained in comprehensive and authoritative volumes. The tablets themselves rest in museums and collections in many countries; sometimes different bits of the same tablet are in different museums. An unbroken tablet—there are only a few of these—is about as big as a hand and is made of clay which is usually unbaked. The writing is called *cuneiform*, i.e. wedgeshaped, for the signs are made up of single wedgeshaped marks which were impressed with a stylus upon the tablet while it was still wet. Most of these tablets date from a couple of centuries around 1700 B.C. and the rest from the last three centuries B.C. (there is yet no satisfactory explanation for the long pause between these two groups). The age of a mathematical tablet must be inferred from the stratum of the mound where it was found, or from the style of the handwriting, for the contents of the tablets give no clue to their age. It seems curious to us, who are familiar with the explosive evolution of mathematics and the sciences during the last couple of centuries, that Babylonian mathematics not only retained its character for almost 2000 years, through violent political changes, but also kept its content within the same boundaries. We cannot, in the available texts, trace any development at all (there are, however, some very old tablets exhibiting an early stage of the Babylonian number system, and one may detect a preference for more elaborate numerical examples in the late texts). It seems, therefore, that the creation of Babylonian mathematics happened with great swiftness and that this short period of rapid growth was followed by a long time span of stagnation. Of the creators of Babylonian mathematics we know nothing whatsoever except the result of their work.

1.2 The Babylonian Number System. A Multiplication Table

Before we can approach Babylonian mathematics we must acquaint ourselves with the Babylonian number system; for it had, as we shall see, an all-pervading influence on the nature of Babylonian mathematics. In the following two paragraphs I shall try to show how it was possible to uncover the structure of this number system from the texts alone, with no previous knowledge. It is, of course, easier to do this when one knows the final result, so the reader must be warned against under-estimating the difficulties that faced the patient scholars who first mapped the road we shall follow.

Figure 1.1 is a copy of the front (obverse) and back (reverse) of an Old-Babylonian tablet. On each side the writing consists of simple signs in two columns denoted on Figure 1.1 by Col. I (on the left) and Col. II; if we count both sides, each column has 24 lines, but for the time being we shall disregard the last line.

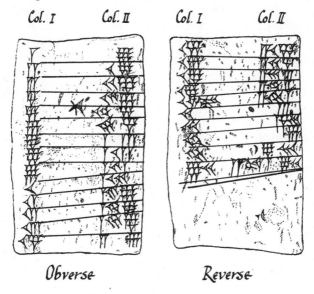

Figure 1.1

Let us consider Column I, beginning at the top. The first entry is a vertical wedge, the second two vertical wedges, and the third three. It is natural to read these as 1, 2, 3. As a matter of fact, the next six lines can equally easily be read as 4, 5, 6, 7, 8, 9, for these are, respectively, the number of vertical wedges in them. We observe, however, that they are grouped in threes—this makes, by the way, the reading of them at a glance simpler—so that, e.g., 8 is written in three levels, two with three wedges in each and one with two. After 9 we find a new sign, a corner-wedge. If this is read as 10, the following eight lines offer no further difficulty, for they are composed of a cornerwedge and our already de-ciphered signs for 1 to 8. They can therefore immediately be read as 11, 12, 13, ···, 18. We shall not bother about the next line. (In fact, it has a special sign for 19 and some erasure marks—but 19 is usually

written as a cornerwedge and nine vertical wedges.) The subsequent four
lines have two, three, four, and five cornerwedges which ought to mean
20, 30, 40, and 50.

To summarize what we have learned so far: the Babylonian numbers
are built up of two basic signs, a vertical wedge meaning 1, and a corner-
wedge meaning 10. The first column lists simply all the integers up to
and including 20, followed by 30, 40, 50.

Let us now apply this knowledge to Column II. Without much trouble
the first six lines can be made out to be 9, 18, 27, 36, 45, 54. We can
now make a strong hypothesis about our text, namely that it is a multiplication table for 9. The seventh and eighth lines should then be 63
and 72, but we find a vertical wedge followed by a 3 and a 12, respectively. Obviously it will not do to read this vertical wedge as 1. The
only thing that makes sense is to let it mean 60. We transcribe these
lines as 1,3 and 1,12, and letting the first 1 mean 60 we have:

$$1,3 = 1 \cdot 60 + 3 = 63 \quad \text{and} \quad 1,12 = 1 \cdot 60 + 12 = 72.$$

The next lines can be transcribed and interpreted as

$$1,21 = 81$$
$$1,30 = 90$$
$$1,39 = 99$$
$$1,48 = 108$$
$$1,57 = 117,$$

all in agreement with our hypothesis that this text is a multiplication
table for 9. The fourteenth line has two vertical wedges and a 6, or
what we transcribe as 2,6. This ought to be $14 \cdot 9 = 126$, and so we
must let the initial 2 mean $120 = 2 \cdot 60$. We can now write the following
lines thus:

$$2,15 = 2 \cdot 60 + 15 = 135$$
$$2,24 = 144$$
$$2,33 = 153$$
$$2,42 = 162$$
$$2,51 = 171,$$

but the next line has only a 3 which must mean 180. If this 3 had
been followed by a sign for zero, or what we would transcribe as 3,0, it
would have been in perfect agreement with what we have found hitherto,
for 3,0 would be $3 \cdot 60 + 0$ just as 2,15 was $2 \cdot 60 + 15 = 135$. We
can only assume, then, that the Babylonians did not use a sign for zero
at the end of a number but left it to the reader to guess that an extra
empty place was intended. We can test this assumption two lines farther

down where we find, opposite 40, a 6 which, when read as 6,0 or
$6 \cdot 60 + 0$ is indeed $40 \cdot 9$. The remaining two unexplained lines opposite
30 and 50 —we are still disregarding the last one—are now trivially
read as

$$4{,}30 = 4 \cdot 60 + 30 = 270 = 9 \cdot 30$$
$$7{,}30 = 7 \cdot 60 + 30 = 450 = 9 \cdot 50.$$

Thus we have seen that the text makes perfect sense if we assume that
the number signs, or digits,† change their value with their place in such
a fashion that to move a digit one place to the left means to multiply its
value by 60.

If one analyzes other texts in the same way as we did the multiplication
table for 9, this assumption is amply justified. We have, then, 59 digits
transcribed as 1, 2, 3, $\cdots$, 59 and written as combinations of cornerwedges and vertical wedges. By means of these digits *all* numbers are
written; this is achieved by assigning importance to the place where a
digit occurs, so that for every place a digit is moved to the left its value
becomes sixty times as large. In transcribing Babylonian numbers we
shall in general, as above, separate the digits by commas. Thus a number
which we transcribe as 1,25,30 can mean

$$1 \cdot 60^2 + 25 \cdot 60 + 30 = 3600 + 1500 + 30 = 5130.$$

But, as we said above in connection with $20 \cdot 9$ and $40 \cdot 9$, it may well
be that we ought to transcribe the number as 1,25,30,0 or, indeed,
1,25,30,0,0 in which cases the values would be 60 or 60^2 times larger
than 5130, for

$$1{,}25{,}30{,}0 = 1 \cdot 60^3 + 25 \cdot 60^2 + 30 \cdot 60 + 0 = 60 \cdot 5130$$

and

$$1{,}25{,}30{,}0{,}0 = 1 \cdot 60^4 + 25 \cdot 60^3 + 30 \cdot 60^2 + 0 \cdot 60 + 0 = 60^2 \cdot 5130.$$

In fact, if nothing else is known, the sequence of digits 1,25,30 can stand
for any one of the numbers $5130 \cdot 60^n$, $n = 0, 1, 2, 3, \cdots$; which of these
is intended must be determined from the context. This is not as formidable
a flaw in the number system as one might believe at first; usually there
is no doubt at all about the proper value.

The Babylonians did, indeed, occasionally use a sign for zero‡ in very
late texts, but only to denote an empty space *inside* a number so as to

† To avoid confusion it must be emphasized that in this context, e.g., 2,24 is called
a *two* digit number; the first digit is 2 and the second 24.

‡ It looked somewhat like ⧨ and is otherwise used as a separation sign, or period.

distinguish, for example, $1,0,30 = 3630$ from $1,30 = 90$. In older texts one simply left an open space between the 1 and the 30 or, even more simply, did nothing at all.

For the sake of completeness, the last line in our multiplication table for 9 says

$$8,20 \text{ times } 1 \text{ is } 8,20$$

and is what we call a *catch line*. Our text is one of a series and the catch line is the first line of the next text.

1.3   The Babylonian Number System. A Table of Reciprocals

What we learned above from the multiplication table for 9 is, however, only part of the tale of the Babylonian number system. The rest can be gleaned from a text which is transcribed, except for its first line, in Figure 1.2. This is a type of which we have found many examples (from both Old-Babylonian and Seleucid† times) differing only in their first lines so that they all contain the numbers in Figure 1.2. The tablet reproduced in Figure 1.3 contains, among other things, two copies of this type of text.

Col. I	Col. II	Col. I	Col. II	Col. I	Col. II
2	30	16	3,45	45	1,20
3	20	18	3,20	48	1,15
4	15	20	3	50	1,12
5	12	24	2,30	54	1, 6 ,40
6	10	25	2,24	1	1
8	7,30	27	2,13,20	1,4	56,15
9	6,40	30	2	1,12	50
10	6	32	1,52,30	1,15	48
12	5	36	1,40	1,20	45
15	4	40	1,30	1,21	44,26,40

Figure 1.2

Like our first text, the one in Figure 1.2 consists of numbers arranged in two columns, Col. I and Col. II. The structure of this table becomes

† The Seleucid Era began in 312 B.C. in Mesopotamia. It is named after Seleucus Nicator, a former general of Alexander's who managed to become king of the Eastern (and greater) part of Alexander's empire, including Mesopotamia.

evident when we form, line by line, the product of the number in Col. I and its mate in Col. II, interpreting them as we learned in the preceding paragraph, thus:

$$2 \cdot 30 \;=\; 60 \;=\; 1,0$$
$$3 \cdot 20 \;=\; 60 \;=\; 1,0$$
$$4 \cdot 15 \;=\; 60 \;=\; 1,0$$
$$5 \cdot 12 \;=\; 60 \;=\; 1,0$$
$$6 \cdot 10 \;=\; 60 \;=\; 1,0$$
$$8 \cdot 7,30 \;=\; 60^2 \;=\; 1,0,0$$

$\cdots$

It appears throughout all thirty lines that the result is always some positive power of 60, and the selection of numbers in Col. I (or at least those less than 60) is explained when we observe that Col. I contains precisely all the integers less than 60 which are factors of some power of 60. Incidentally, these are best characterized in the following fashion: The prime factorization of 60 is

$$60 \;=\; 2^2 \cdot 3 \cdot 5,$$

and hence that of a power of 60 is

$$60^n \;=\; 2^{2n} \cdot 3^n \cdot 5^n.$$

If an integer contains a prime factor other than 2, 3, and 5, it cannot divide a power of 60.[†] On the other hand, if it contains no prime factors other than 2, 3, or 5, we can surely find a power of 60 which it will divide. The following representative example will make this point clear. Take

$$24 \;=\; 2^3 \cdot 3;$$

in order to bring the right side into the form $2^{2n} \cdot 3^n \cdot 5^n$ we multiply it by $2 \cdot 3 \cdot 5^2 = 150$, and thus we have

$$24 \cdot 150 \;=\; 2^4 \cdot 3^2 \cdot 5^2 \;=\; 60^2$$

or, in Babylonian notation,

$$24 \cdot 2,30 \;=\; 1,0,0$$

in agreement with the text in Figure 1.2.

Thus the absence from Col. I of 7, 11, 13, 17, 19, 23, etc. and their multiples is explained.

[†] We are here making use of the theorem that every positive integer has one, and only one, prime factorization. For a proof, see e.g. I. Niven, *Numbers: Rational and Irrational*, NML vol. 1, Appendix B, p. 117.

Figure 1.3. A photograph of a tablet from Nippur (SE. of Babylon). A heavy vertical score divides it into two parts. To the left the teacher or, I suspect, an older boy has written a multiplication table for 45 (the last seven or eight lines are broken off), and to the right a beginner has tried to copy it. One need not be an expert on cuneiform writing to see how helpless and unsteady the little fellow's hand was. He did not even finish his work.

The table to the left is transcribed below (the square brackets in the transcription simply mean that I have restored what is inside them); the cluster of wedges meaning *a-rá* (times) ends with a vertical wedge which in the last lines might be mistaken for 1. The second line has an error (2,30 for 1,30).

45 *a-rá* 1	45	[*a-rá* 6]	4,30	[*a-*]*rá* 12	9
[*a-rá* 2]	2,30	[*a-rá* 7]	5,15	[*a-*]*rá* 13	9,45
[*a-rá*] 3	2,15	[*a-rá*] 8	6	[*a-*]*rá* 14	10,30
[*a-rá* 4]	3	[*a-rá*] 9	6,45	[*a-*]*rá* 15	11,15
[*a-rá* 5]	3,45	[*a-rá*] 10	7,30	[*a-*]*rá* 16	1[2]
		[*a-r*]*á* 11	8,15		

(a)

(b)

(c)

Figure 1.4. Ancient Mesopotamian schoolrooms (Palace at Mari). (a) shows two adjoining schoolrooms, (b) the better preserved one which, in (c), is occupied by workers of the expedition.

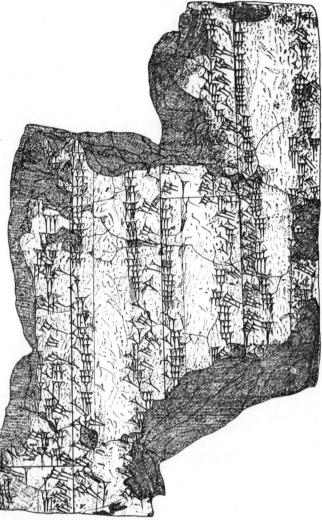

Figure 1.5. The tablet copied here is the reverse of Figure 1.3. It is doubtless the work of a schoolboy, but one more advanced than the little fellow who wrote on the obverse. On this side are several copies of a multiplication table and of a standard reciprocal table.

In studying our table, we found that if we read the numbers as in the preceding paragraph, the two columns have entries whose products are various powers of 60. But we also learned that any positive power of 60 was written by the Babylonians as 1. It is, in fact, this seeming weakness of the Babylonian number system that gives the key to the correct understanding of our table, for it suggests as a simpler interpretation that the products of the numbers in Col. I and their partners in Col. II always are 1 rather than different powers of 60 or, in other words, that Col. II gives the reciprocals of the numbers in Col. I. This interpretation requires, of course, that we no longer read the entries in Col. II as integers, but as integers divided by an appropriate power of 60. For example, we used to read the 7,30 in the sixth line as the integer $7 \cdot 60 + 30 = 450$ and found that multiplied by 8 it gave 1,0,0 or 60^2. If 7,30 now is to represent 1/8, its new value must be 60^2 times smaller. In other words, instead of interpreting line six of Col. II, Figure 1.2 as

$$7,30 \ = \ 7 \cdot 60 + 30 \ = \ \frac{60^2}{8} \, ,$$

we now interpret it as

$$\frac{7,30}{60^2} \ = \ \frac{1}{8}$$

or

$$\frac{7 \cdot 60 + 30}{60^2} \ = \ \frac{7}{60} + \frac{30}{60^2}$$

which is, indeed, 1/8. So now the 7 in 7,30 stands for 7/60 and the 30 for $30/60^2$.

Similarly, the number 44,26,40 facing 1,21 (=81) in the last line has to be read as

$$\frac{44}{60^2} + \frac{26}{60^3} + \frac{40}{60^4}$$

to equal 1/81.

We see again that moving a digit one place to the left increases its value by a factor 60 or, what is the same, that moving a digit one place to the right divides its value by 60 and that this principle is carried *even beyond the units' place*. This last is the important new thing, for it means that certain fractions can be written simply in the Babylonian number system.

But this also implies that a sequence of digits such as 1,25,30 which we once read as

$$1 \cdot 60^2 + 25 \cdot 60 + 30 \ = \ 5130$$

now may mean 5130 times 60^k, where k is any whole number, positive, negative, or zero. Whenever we are sure of where the units' place is—as we most often are from the context—we shall, in our transcription of the number, separate the whole from the fractional part by a semicolon so that, for example,

$$1,25;30 \ = \ 1 \cdot 60 + 25 + \frac{30}{60} \ = \ 85\tfrac{1}{2} \, ,$$

$$1;25,30 \ = \ 1 + \frac{25}{60} + \frac{30}{60^2} \ = \ 1\tfrac{17}{40} \, ,$$

and so on. It must be emphasized once more, however, that neither semicolons nor zeros at the end are in the original texts, but are added by us in our modern transcriptions for the sake of clarity.

Problems

1.1 Verify that 44,26,40, with the above interpretation, is the reciprocal of 81. Where should we place the semicolon?

1.2 Identify the multiplication table and the table of reciprocals in Figure 1.5. (The table of reciprocals begins with the statement that $\tfrac{2}{3}$ of 1 is 0;40.) Try to find the student's errors.

1.4   Positional Number Systems

The similarities between our own number system and that of the Babylonians are several: we, as they, employ a finite number of symbols or digits (we use ten) to express all integers; and we, too, make them do the job by assigning importance to the position of a digit, so that for every place it is moved to the left its value is multiplied by a constant factor (with us 10, with the Babylonians 60). We, as they, make use of an extension of this principle to express certain fractions (decimal fractions in our case) by carrying even beyond the units' place the rule that moving a digit one place to the right means to divide its value by the constant factor 10, or 60. Incidentally, the numbers 10 and 60 that play such a crucial role are called the *bases* for the two number systems which are called the *decimal* and the *sexagesimal* system, respectively; and just as we speak of *decimal fractions* we call their Babylonian counterparts *sexagesimal fractions*.

The differences between the two systems, viz. the unfamiliar Babylonian base 60 and the absence of the equivalent of the decimal point in the sexagesimal system, are perhaps at first sight more conspicuous than the similarities, but they are really less important. In order to make this clear it is well to consider the problem of number notation in more general terms.

There is, of course, nothing particularly outstanding about the numbers 10 and 60; our predecessors' choice of 10 is simply a matter of biological whim, and though the Babylonians were not above counting on their fingers, as we can infer from their special sign for 10, their choice of 60 as a base also had its motive outside mathematics, as we shall discuss later in this section. It is, in fact, not too difficult to prove that any integer b greater than 1 can serve as the base of a *positional* or *place-value* number system, as we call a number system with the common characteristics of the decimal and the sexagesimal number systems. In such a system we shall need b different symbols or digits whose principal values are $0, 1, 2, \cdots, b - 1$. To move a digit one place to the left will mean to multiply its value by b and to move it one place to the right, even beyond the units' place, will mean to divide its value by b.

We shall illustrate this by an example which, as it happens, has recently become of practical importance in electronic computers, namely the *binary system*, where b is two. We then have two digits, 0 and 1. The first ten whole numbers are written in this system thus:

$$1, \quad 10, \quad 11, \quad 100, \quad 101, \quad 110, \quad 111, \quad 1000, \quad 1001, \quad 1010.$$

In order to translate the binary number 1001011 into decimal notation we observe that

$$1001011 = 1\cdot2^6 + 0\cdot2^5 + 0\cdot2^4 + 1\cdot2^3 + 0\cdot2^2 + 1\cdot2 + 1 = 75.$$

Conversely, if we are to write, e.g., the number 308 (decimally written) in binary form, we note that 308 lies between the following two consecutive powers of 2:

$$2^8 = 256 \quad \text{and} \quad 2^9 = 512,$$

and so

$$308 = 2^8 + 52.$$

52 is between

$$2^5 = 32 \quad \text{and} \quad 2^6 = 64,$$

so

$$52 = 2^5 + 20.$$

Similarly,

$$20 = 2^4 + 4 = 2^4 + 2^2,$$

and so

$$308 = 2^8 + 2^5 + 2^4 + 2^2$$
$$= 1 \cdot 2^8 + 0 \cdot 2^7 + 0 \cdot 2^6 + 1 \cdot 2^5 + 1 \cdot 2^4 + 0 \cdot 2^3 + 1 \cdot 2^2 + 0 \cdot 2 + 0;$$

this, written in binary form, becomes 100110100.

The positional number notation lends itself particularly well to arithmetical computations, and herein lies its importance. All one has to know once and for all are the tables that give the products and sums of any two one-digit numbers; the rest is carried out according to the familiar computation schemes taught in elementary school.

To return to our binary example, the multiplication and addition tables are as simple as can be, namely:

·	0	1		+	0	1
0	0	0		0	0	1
1	0	1		1	1	10.

Accordingly, a binary multiplication is carried out thus:

```
        1 1 0 1
          1 1 0
        ───────
        0 0 0 0
        1 1 0 1
      1 1 0 1
    ─────────────
    1 0 0 1 1 1 0.
```

Problem

1.3 Check this multiplication by translating it into decimal notation.

That the binary system is in vogue in the computer world of late is due to two of its features: it uses only two digits, and this agrees well with the two things that a light bulb can do—be on or off; and its tables for addition and multiplication are easy to teach to a machine. The price one pays for this simplicity is the length even of rather modest numbers; for example, $1024 = 2^{10}$ already requires eleven digits.

We can now return to the conspicuous differences between the sexagesimal and the decimal systems; it should be apparent that the base 60, though unfamiliar, does as well as 10. To be sure, each base has its advantages and disadvantages; an obvious disadvantage of the larger base 60 is that a multiplication table has a size (59 by 59) that practically prohibits memorization; on the other hand it is possible to write rather large numbers with few sexagesimal digits.

One further advantage of the Babylonian base is that more fractions can be written as finite sexagesimal fractions than can be written as finite decimal fractions. We have, in fact, already described such fractions in the section on the reciprocal table, but it may be natural to ask the more general question:

When does a *reduced* fraction (i.e. a fraction in lowest terms) p/q have a finite expansion in a number system with the base b?

We remark first that a finite decimal fraction can be thought of as a fraction whose denominator is a power of 10, and a finite sexagesimal fraction as one whose denominator is a power of 60. Similarly, a finite fraction in any other number system with the base b is a fraction whose denominator is a power of b. Our question is then: when can a reduced fraction p/q be turned into a fraction with denominator b^n? Since we can change the denominator of a reduced fraction only by multiplying both the numerator and denominator of the fraction by some integer, the answer is: p/q can be turned into a fraction p'/b^n precisely if the denominator q contains only prime factors which also appear in b^n and therefore in b.

So since 2 is itself a prime, the only reduced fractions that can be written as finite binary fractions are those whose denominators are already powers of 2. Those which can be turned into finite decimal fractions are the ones whose denominators have no other prime factors than 2 and 5 since $10 = 2 \cdot 5$. But since $60 = 2^2 \cdot 3 \cdot 5$ the permissible prime factors for finite sexagesimal expansions are 2, 3, and 5. Thus if we consider the denominators 2, 3, 4, $\cdots$, 20, only four of them will yield finite binary fractions, seven will give finite decimal equivalents, while thirteen have finite sexagesimal expansions.

Problem

1.4 Of the numbers 2, 3, 4, $\cdots$, 20, which have reciprocals with finite binary expansions, which have reciprocals with finite decimal expansions, and which have reciprocals with finite sexagesimal expansions?

The other major difference, namely the absence of the equivalent of the decimal point, is, to be sure, a flaw in the sexagesimal system. Yet it is not as serious as one might think at first. We need only remember that when we are concerned with multiplication and division of decimal fractions the first thing we do is to forget about the decimal points;

after all, they have no influence on the sequence of digits in the result, but govern solely its size. In fact, when we use a slide rule or look up the logarithm of a number we are in a situation not too different from that of the Babylonians, because we obtain first the digits of the answer and have then to decide the position of the decimal point. At any rate, this deficiency is a small price to pay for the enormous advantage that operations with fractions are usually no more complicated than those with integers.

The origin of the sexagesimal system cannot be established with certainty. The following is a much simplified account of a plausible explanation. We know that there were in early times several systems of units of weight and measure, among them one where the larger unit was 60 times as large as the smaller. It was customary to write a measure of, say, seventy-two smaller units as a large 1 followed by a small 12; this represented one large and twelve small units. But this idea, that the number of large units was written with a large character, was also used where the ratio of large unit to small unit was different from 60 (e.g., in early texts 100 is sometimes written as a large 10). Each of these different schemes had in it the germ of a positional system; for, after a while, the large characters would tend to be written in ordinary size, and then all that is needed is some one with the bright idea of boldly extending the positional principle to several places. That it was the base 60 that came to play the important role may have had to do with the fact that the principal unit of weight for silver—the *mana*—was subdivided into 60 *shekels*. This may have given the impetus to the consideration of sixtieths as natural subdivisions of units, on the one hand— whence sexagesimal fractions—and on the other, the preference for the base 60 in general.

It should be added that an entirely consistent use of the sexagesimal system is to be found only in the mathematical and astronomical texts, and even in astronomical texts one can find year numbers written as, e.g., 1–*me* 15 (meaning 1 *hundred* 15) instead of 1,55. In practical life the Babylonians showed the same profound disregard for rationality in their use of units for weight and measure as does the modern English-speaking world.

1.5 Babylonian Arithmetic

We saw in the preceding paragraph that a disadvantage of a base as large as 60 is the uncomfortably large size of the multiplication table showing the products of any two one-digit numbers. One may well shudder to imagine poor Babylonian schoolboys trying to commit such a

59 by 59 table to memory, and is therefore much relieved to find that there were quantities of tables of various kinds including multiplication tables; so it is clear that such memorization was unnecessary.

This is not to say that we have tablets containing the 59 times 59 products, for we do not. What we find are many tables of the type discussed in Section 1.2—that was the 9-table—arranged according to multiples of p:

1	p
2	$2p$
3	$3p$
.	.
.	.
.	.
19	$19p$
20	$20p$
30	$30p$
40	$40p$
50	$50p$,

and sometimes ending with p^2. We call p the *principal number* of the multiplication table. From this table any multiple of p can readily be found—$47p$ is simply the sum of $40p$ and $7p$ both of which are tabulated.

One might now expect that there were fifty-nine such tables with $p = 1, 2, 3, \cdots, 59$. But what we actually find is a selection of principal numbers which is, at first, quite puzzling. We have, for example, a multiplication table with $p = 44,26,40$, an enormous number, but none for $p = 17$. It is the presence of this curious principal number 44,26,40 that makes the puzzle pieces fall into place, for 44,26,40 is the last number in the standard reciprocal table discussed in Section 1.3. It seems that the principal numbers are essentially the numbers we find in a standard reciprocal table. One exception is 7 which, quite naturally, appears as a principal number though it is absent from the reciprocal tables, and so the product of any two numbers can quite easily be found. The virtual coincidence of principal numbers and the numbers in the reciprocal table gives us a clue to the ways in which the Babylonians actually computed. It is quite clear that the reciprocal table combined with these multiplication tables also served for divisions, since a divided by b is a multiplied by the reciprocal of b, or

$$\frac{a}{b} = a \cdot \frac{1}{b}.$$

In our decimal system we have a variety of rules and short-cuts which facilitate computations, such as: to multiply by 5 we divide by 2 and multiply by 10; a number is divisible by 3 (or 9) if the sum of the digits is divisible by 3 (or 9). If one works consistently with the sexagesimal system one soon finds many such simple devices; that so many more rules are possible in the sexagesimal than in the decimal system is because the base 60 has so many divisors.

Problem

1.5 Try to discover and formulate rules for multiplying by 6 and for dividing by 6 in the sexagesimal system.

Sexagesimal calculations were further assisted by quite a large variety of tables. We find extended reciprocal tables—even giving the reciprocals to several places—of numbers such as 7 and 11 whose reciprocals do not have finite sexagesimal expansions. There are tables for the computation of compound interest (at exorbitant rates, e.g., 20% yearly), tables of squares and square roots, of cubes, and several quite complicated tables which indicate an interest in numerical procedures far beyond the requirements of simple arithmetic.

Thus it is perfectly clear that the Babylonians found no more difficulties in arithmetical computations than we do today. In this respect they were unique in the classical world, and it is therefore not surprising that when Greek astronomy had reached the stage where extensive calculations were called for, the Greek astronomers—as we shall see later with Ptolemy—turned to the sexagesimal number system for a sensible way of expressing fractions. This is the reason why a special use of Babylonian fractions has been kept continuously alive and is used even now, for example, in the subdivision of degrees and hours, the units for the measurement of the two basic quantities observed in classical astronomy, namely angle and time. Inconsistently, the Greeks wrote the whole part of the measure in their own system, and so do we when we write, e.g., 120°12′20″. Few people realize that when we say it is 2 hours 30 minutes and 10 seconds† p.m. we are actually talking the language of the Babylonians of 4000 years ago, who, somewhat more simply, would have said that 2;30,10 hours have passed since noon.‡

† The words *minute* and *second* come from the Latin designations of these parts, *pars minuta prima* (small part of the first kind) and *pars minuta secunda* (small part of the second kind), respectively.

‡ To prevent misunderstanding it should be said that, though our degrees were used in Seleucid astronomy, our hours are of Egyptian origin.

1.6 Three Babylonian Mathematical Texts

To present an exhaustive picture of Babylonian mathematics is clearly beyond the scope of this book, but the following three texts may give the reader some feeling for its nature. The first two texts are Old-Babylonian, i.e. from the centuries about 1700 B.C., and the last is Seleucid, i.e. from the last three centuries B.C. As already mentioned, the dates cannot be inferred from the content, and in these cases the style of writing yields the dating.

The mathematical cuneiform texts are usually divided into two classes, table texts and problem texts, although the line of demarcation between the two types is by no means sharp. The multiplication table for 9 and the reciprocal table are perfect examples of table texts. A problem text often consists of many problems of a similar kind, pedagogically arranged in order of increasing difficulty. The examples in A and C below are selected from two such texts with many sections.

A. *Quadratic equations.* We shall here present in translation the sixth and seventh sections of a tablet† with twenty-four sections. The tablet is Old-Babylonian. The semicolons are added in transcription, but we shall see that there is no possible doubt whatever as to their positions.

(1)   I have added the area and two thirds of the side of my square and it is 0;35. You take 1, the "coefficient". Two thirds of 1, the coefficient, is 0;40. Half of this, 0;20, you multiply by 0;20 (and the result) 0;6,40 you add to 0;35, and (the result) 0;41,40 has 0;50 as its square root. 0;20, which you multiplied by itself, you subtract from 0;50 and 0;30 is the (side of) the square.

This example states and solves the quadratic equation

$$x^2 + \tfrac{2}{3}x \;=\; 0;35.$$

$\tfrac{2}{3}$ is written with a special sign, and the first step is to convert $\tfrac{2}{3}$ to its sexagesimal equivalent 0;40. If we follow the solution step by step, we are led to what we would express as

$$x \;=\; \sqrt{\left(\frac{0;40}{2}\right)^2 + 0;35} \;-\; \frac{0;40}{2} \;=\; 0;30 \,.$$

Indeed, the positive solution of

$$x^2 + px \;=\; q$$

† BM 13901 (BM stands for British Museum) published in O. Neugebauer's *Mathematische Keilschrifttexte*, vol. III, Berlin 1937.

is, according to the quadratic formula,

$$x = \sqrt{\left(\frac{p}{2}\right)^2 + q} - \frac{p}{2}.$$

Problem

1.6 Show that this expression is equivalent to one of the solutions of $x^2 + px = q$ yielded by the quadratic formula.

(2)   I have added seven times the side of my square to eleven times its area and it is 6;15. You take 7 and 11. You multiply 11 by 6;15 and it is 1,8;45. You halve 7 (and obtain 3;30). You multiply 3;30 and 3;30. You add (the result) 12;15 to 1,8;45 and (the result) 1,21 has 9 as its square root. You subtract 3;30, which you multiplied by itself, from 9 and you have 5;30.

    The reciprocal of 11 does not divide. What shall I multiply by 11 so that 5;30 results? 0;30 is its factor. 0;30 is the (side of) the square.

Here we see a solution of the quadratic equation

$$11x^2 + 7x = 6;15.$$

The multiplication by 11 has the effect of turning this into a quadratic equation in $11x$:

$$(11x)^2 + 7 \cdot (11x) = 6;15 \cdot 11 = 1,8;45,$$

where the coefficient of the square term is 1.

    The equation

$$u^2 + 7 \cdot u = 1,8;45$$

is now solved in the same manner as example A(1); thus

$$u = \sqrt{\left(\frac{7}{2}\right)^2 + 1,8;45} - \frac{7}{2} = 5;30 ,$$

and from

$$11x = u = 5;30$$

we get

$$x = 0;30.$$

This last division by 11 cannot be carried out in the usual manner, viz. by multiplying by the reciprocal of 11, because "the reciprocal of 11 does not divide", i.e. it has no finite sexagesimal equivalent.

We note first that these two problems are anything but practical problems. That we are asked to add areas and lengths shows clearly that no real geometrical situation is envisaged. In fact, the term "square" has no more geometrical connotation than it does in our algebra.

We observe further, that there is quoted no general formula or theorem such as our quadratic formula, which once and for all solves all quadratic equations, and this holds in general for all of Babylonian mathematics. None the less, the instructions are so specific that we are pretty sure of the general procedure; and after one has worked through a large number of problems, no doubt can remain.

There is, however, no explicit indication anywhere in the mathematical texts of how these implied rules were found. It is not until the time of the Greeks that the notion of a proof of a theorem assumes the central role which it has held ever since in mathematics.

B. *The Diagonal of a Square.* Figure 1.6b is a copy and Figure 1.6c a transcription of the small Old-Babylonian tablet† pictured in Figure 1.6a (the photograph is full size).

We find three numbers:

$$a = 30,$$
$$b = 1,24,51,10,$$
$$c = 42,25,35.$$

We observe first that

$$c = a \cdot b$$

if we insert semicolons in the proper places, for to multiply by 30 is essentially the same as to divide by 2. If a is taken to mean the side of a square, as suggested by the figure, and c the diagonal, then, by Pythagoras' theorem, $c^2 = 2a^2$, and $c = a\sqrt{2}$, so b ought to be an approximation of $\sqrt{2}$, that is, if it is interpreted as $1;24,51,10$. This is indeed correct; for

$$(1;24,51,10)^2 = 1;59,59,59,38,1,40$$

is very close to 2.

Thus what the tablet tells us is that if the side of a square is $a = 30$ then its diagonal is $c = 42;25,35$, and it also gives an excellent approximation to the proper factor $\sqrt{2}$.

---

† YBC 7289 (YBC = Yale Babylonian Collection) published in O. Neugebauer and A. Sachs [5], p. 42 ff. Here and throughout, the number in brackets refers to the listing in the bibliography, pages 131–133.

Figure 1.6a

So we learn from this simple tablet with but one figure and three numbers on it that the Babylonians knew that the diagonal of a square is $\sqrt{2}$
times its side; this implies that they were acquainted with at least a
special case of what we are wont to call the Pythagorean theorem—
this was some 1200 years before Pythagoras is supposed to have lived—
and from other texts (e.g. C below) we can see that they did indeed
make use of this theorem in its full generality. Moreover, we learn that
the Babylonians possessed arithmetical techniques sufficient to achieve a
fine approximation to $\sqrt{2}$.

C. *The Area of a Trapezoid.* The following is the third of seven preserved sections of a tablet†—the reverse is destroyed—from the Seleucid
period.

† VAT 7848 (VAT = Vorderasiatische Abteilung, Tontafeln, Staatliche Museen,
Berlin), published in O. Neugebauer and A. Sachs [5], p. 141 ff.

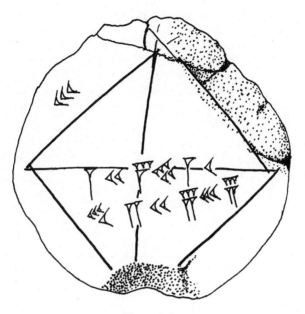

Figure 1.6b

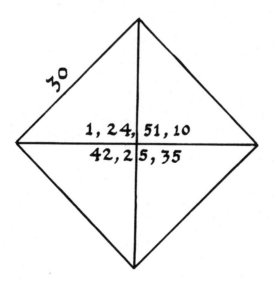

Figure 1.6c

A trapezoid 30 is the length, 30 the second length, 50 the upper width, 14 the lower width. 30 times 30 is 15,0. Subtract 14 from 50 and the remainder is 36. Half of it is 18. 18 times 18 is 5,24. Subtract 5,24 from 15,0 and the remainder is 9,36. What should I multiply by itself so that the result will be 9,36? 24 times 24 is 9,36. 24 is the dividing line. Add 50 and 14, the widths, and (the result is) 1,4. Half of it is 32. Multiply 24, the dividing line, by 32, and (the result is) 12,48 . . . .

The rest of the section is devoted to a change in measuring units.

This example is concerned with finding the area of an isosceles trapezoid of dimensions (see Figure 1.7)

$$l = 30, \qquad w_1 = 50, \qquad w_2 = 14.$$

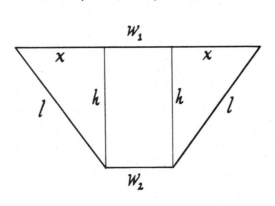

Figure 1.7

The first step is to find what in the figure is called x:

$$x = \frac{w_1 - w_2}{2} = \frac{50 - 14}{2} = 18.$$

Next the altitude h—the "dividing line" of the text—is determined from the Pythagorean theorem:

$$h = \sqrt{l^2 - x^2} = \sqrt{15,0 - 5,24} = \sqrt{9,36} = 24.$$

Finally the area is computed according to the correct formula:

$$A = h \cdot \frac{w_1 + w_2}{2} = 24 \cdot 32 = 12,48.$$

1.7 Summary

From the multitude of texts Babylonian mathematics emerges as a creation with well-defined features, and of these the reader has already had several glimpses. The backbone of the structure is the sexagesimal place-value system which made the Babylonian mathematicians sovereign computers, and it is therefore hardly surprising that Babylonian mathematics shows a strong preference for what we today would call algebra and number theory. Though there is a considerable amount of geometrical knowledge, geometry often serves merely as a guise for essentially algebraic problems. We saw this in the examples A in Section 1.6, where areas and lengths are added in violation of geometrical sense; and whenever a geometrical problem is posed it is for the express purpose of computing some numerical quantity, be it length, area or volume. Furthermore, there are no examples of a theorem, formulated in some generality, although some of the worked examples from which we distill general procedures are done in such detail that the step to a theorem is but a short one. Finally, we never find an explicit proof in Babylonian mathematical texts.

The method of handing down the mathematical knowledge through the generations was therefore entirely different from the one employed in modern mathematics; yet it may not appear so strange to someone who has been subjected to an old-fashioned high school algebra course, where one learned of, say, quadratic equations by doing a large number of problems with varying coefficients instead of stating and proving a theorem which shows once and for all how to solve any quadratic equation that may arise. For, as we have seen, the problem texts are precisely such long lists of problems playing variations on a central theme.

After this rough outline of the general nature of Babylonian mathematics we turn to a brief description of its contents.

In algebra we find the solutions of first and second degree equations. Quadratic equations are often given in the equivalent form of two equations with two unknowns, such as

$$x + y = a, \qquad xy = b,$$

whence one finds immediately that x and y are the solutions of

$$z^2 - az + b = 0.$$

Even special third degree equations of the form

$$x^2(x + 1) = a$$

were solved. The problems are often worded in such a way that, when they are translated into modern algebraic letter-notation, extremely complicated expressions arise, with parentheses within parentheses, and one cannot help feeling most impressed with the skill of the Babylonians who were able to reduce such expressions to standard forms of equations without the aid of our algebraic techniques. Incidentally, we find several instances of problem texts in each of which the different problems always yield the same answer; this is a parallel to the modern habit of giving answers in the back of the book.

Of geometrical knowledge we recognize first of all an unrestricted use of the so-called Pythagorean theorem, so its discovery antedates Pythagoras by a millenium and a half. There are furthermore the correct area formulas for simple geometric figures such as triangles and trapezoids (see example C, Section 1.6), and poor approximations of the area and perimeter of a circle (using $\pi \sim 3$), although some recently found texts now in the process of publication indicate the use of better values for π. And we find formulas, some correct and others not, for the volumes of various solids.

Let us finally mention a text which, perhaps more than any other, shows the high level of sophistication reached by Babylonian mathematics. It was published in 1945† and has to do with what we call Pythagorean numbers. A triple of whole numbers, such as 3, 4, 5, or 5, 12, 13, denoting the side lengths of a right triangle, is called a *Pythagorean number triple*; in other words, it is a solution in positive *whole* numbers of the equation

$$x^2 + y^2 = z^2.$$

It is natural to ask how many such triples there are, and how one may find them. Of course, from the triple 3, 4, 5 we can immediately form infinitely many other Pythagorean triples, viz. $3n$, $4n$, $5n$, where $n = 2, 3, 4, \cdots$, but this is so obvious that we count them all as one, represented by 3, 4, 5, and called a *reduced* triple. There is a theorem which states that:

If p and q take on all whole values subject only to the conditions

    1) $p > q > 0$,

    2) *p and q have no common divisor (save 1),*

    3) *p and q are not both odd,*

---

† Plimpton 322 (the tablet is No. 322 in the Plimpton Collection of Columbia University, New York) published in [5], p. 38 ff.

then the expressions

$$x = p^2 - q^2,$$
$$y = 2pq$$
$$z = p^2 + q^2,$$

will produce all reduced Pythagorean number triples, and each triple only once.†

EXAMPLES: $p = 2$, $q = 1$ produce $x = 3$, $y = 4$, $z = 5$;
$p = 3$, $q = 2$ produce $x = 5$, $y = 12$, $z = 13$.

It is clear that the Babylonians knew some form of this theorem, for in the text in question, Plimpton 322, we find listed in fifteen lines corresponding values z^2/y^2, x, z where x, y, z are reduced Pythagorean number triples. The numbers involved are so large (e.g. $x = 3,31,49$, $y = 3,45,0$, $z = 5,9,1$) that trial-and-error methods are excluded. Precisely how the table was constructed and what its purpose was are still matters of controversy; this is so, at least in part, because the left side of the tablet, which doubtless contained important clues, is broken off.

Problems

1.7 Given $x = 100$, how many positive integers y and z can be found so that x, y, z form a reduced Pythagorean triple with

$$x^2 + y^2 = z^2 \, ?$$

Answer the same question for $x = 210$, $x = 420$, $x = 35$.

1.8 Reconstruct the sexagesimal table, a fragment of which is given in Figure 1.8. The table is not a translation of an ancient text, but is most useful for extensive work in the sexagesimal system.

This exercise simulates a common problem facing scholars of ancient texts; however, Fate is usually less generous with what she leaves us.

This is, as stated, a brief summary of the content of Babylonian mathematics. It may be appropriate here to ask what the state of mathematical knowledge was in the other contemporary great culture, the Egyptian one. It so happens that there are preserved two mathematical papyri, the *Rhind Papyrus* and the *Moscow Papyrus*, which give us an idea of the character and content of Egyptian mathematics. It is most disappointing for, contrary to all legends, Egyptian mathematics never

---

† For a proof of this theorem, see e.g. H. Rademacher and O. Toeplitz [6], Section 14, p. 88.

Figure 1.8

managed to get beyond a most elementary stage. The Egyptians' geometrical knowledge was, if we read the texts correctly, about the same
as that of the Babylonians with one important omission: the Pythagorean
theorem. Even the oft repeated statement that the Egyptians knew the
3, 4, 5 right triangle has no basis in available texts, but was invented
some eighty years ago. Outside geometry they did not get past elementary
arithmetic. The explanation of this is doubtless that they had the natural
but unfortunate idea to admit only fractions with numerator 1, i.e.
fractions of the form $1/n$, with one exception, namely 2/3. Thus they
would represent

$$\frac{2}{5} \text{ as } \frac{1}{3}+\frac{1}{15} \quad \text{ or } \quad \frac{9}{10} \text{ as } \frac{2}{3}+\frac{1}{5}+\frac{1}{30},$$

whereas the Babylonians would write 0;24 and 0;54. It is hardly surprising that additions and multiplications, mere trivialities for the Babylonians, remained problems of ultimate complexity for the Egyptians. The Greeks inherited this Egyptian fashion, but when Ptolemy the astronomer is about to embark on serious computations in his *Almagest* he says: "We shall in general make use of the sexagesimal number notation because of the inconvenience of fractions."

The recent re-discovery of Babylonian mathematics not only added directly to our knowledge of the past, but it even forced upon us a reappraisal of Greek mathematics; for we realise now, on the one hand, how greatly the Greek mathematicians were indebted to their Babylonian precursors and, on the other, that what used to be thought of as products of the decay of Greek mathematics is, in fact, a direct continuation of the ancient oriental tradition. However, the old statement that mathematics began with the Greeks is still true, although in a more restricted sense; for it was they who gave a central place to the formulation and proof of theorems, thus giving mathematics the form it has retained ever since.

CHAPTER TWO

Early Greek Mathematics and Euclid's Construction of the Regular Pentagon

2.1 Sources

The problems confronting us when we wish to establish a firm textual basis for the study of Greek mathematics are entirely different from the ones we met in Babylonian mathematics. There our texts—the clay tablets—might be broken or damaged, and the terminology might be obscure and understandable only from the context. But one thing was beyond doubt, and that was the authenticity of the texts, for these were the very tablets the Babylonians themselves had written.

Let us now take Euclid's *Elements*, the work with which we shall be concerned in this chapter, as an example illustrating how different the situation is when we deal with Greek mathematical texts. It was written about 300 B.C., as we shall see, but the earliest manuscripts containing the Greek text date from the tenth century A.D., i.e., they are much closer in time to us than to Euclid.†

Thus even our oldest texts are copies of copies of copies many times removed, and from these we must try to establish what Euclid himself wrote. This is a detective problem of no mean proportions, and classical scholars have developed refined techniques for solving it. The procedure is, in crude outline, as follows:

We compare manuscripts X and Y. If Y has all the errors and peculiarities of X and in addition some of its own, it is a fair assumption that

---

† There are, however, some snippets of Greek papyri from the first centuries A.D. containing parts of the *Elements*; they are so few and so small that they cannot give any idea of the work as a whole, but they provide valuable spot checks.

Y is a copy, or a copy of a copy, of X. If X and Y have a number of errors in common and each some of its own, they are probably both derived from a common archetype Z, which may be lost but reconstructible. In this fashion the extant manuscripts can be arranged in families, each family represented by an archetype. From the archetypes the original text is then reconstructed.

The text critic must possess a number of skills beyond a thorough knowledge of the language and subject of the manuscripts; he must, for example, be familiar with the history of the language and of styles of handwriting and he must have a thorough command of the ancient commentaries of the text.

It is, however, easier to reconstruct a mathematical text than a literary one. After all, one can restore the missing word in:

"In an isosceles triangle the angles at the . . . are equal"

with a much higher degree of certitude than in:

"Rough winds do shake the . . . buds of May".

J. L. Heiberg, the Danish classical scholar who with unbelievable industry gave us the definitive editions of most of the Greek mathematical texts, found that the extant Euclid manuscripts fall into two families. All but one are descendents of an edition by Theon of Alexandria, a busy editor and commentator of the fourth century A.D. One manuscript seems, however, to be derived mainly from a version free of Theon's recensions, but based on a later copy of Euclid than the one Theon had used. Taking these and other facts into account, Heiberg succeeded in establishing as trustworthy a Greek text of Euclid's *Elements* as possible, and it was published between 1883 and 1888. This edition formed the base of all later investigations and translations of Euclid, e.g., T. L. Heath's English version, which is now readily available in paperback editions [8].†

But Euclid's *Elements* was, of course, known in the Western World long before Heiberg's edition. Already in the reign of the Calif Harun ar-Rashid (786–809), whose fame has been sustained by the *Tales of the Arabian Nights*, Euclid was rendered in Arabic by al-Hajjaj, and several Arabic versions followed, some of them drastically abbreviated and others taking great liberties with the Greek original. In the twelfth century some of the Arabic versions were introduced in Europe in Latin translations (Adelard, Gerard of Cremona), and many other Latin

† Here and throughout, the number in brackets refers to the listing in the bibliography, pp. 131–133.

translations appeared in the thirteenth and fourteenth centuries. In 1482 the first printed edition of Euclid (the Campanus translation) was published, and the first Latin translation from the Greek, by Zamberti, appeared in 1505. A Greek text was printed in 1533.

We see here a characteristic pattern: first, translations from the Greek into Arabic in the ninth century, then the Latinization of Arabic versions in the twelfth (the time of the Crusades, during which the contacts between Christians and Muslims were not always bloody), then the printing of the Latin versions towards the end of the fifteenth century, followed closely by Latin translations from the Greek and the Greek text itself (we are now in the Renaissance), and finally, the scholarly definitive edition of the Greek text during the latter half of the nineteenth century.

This could be the history of almost any Greek mathematical, or indeed, technical text, and it illustrates well the tastes and interests of the various periods. There may be variations of different sorts—some of Archimedes' and Apollonius' works, for example, are preserved only in their Arabic versions—and the dates may shift a bit, but in the main the pattern holds.

2.2 Greek Mathematics before Euclid

Our knowledge and esteem of Greek mathematics is based mainly on the extant works of Euclid, Archimedes and Apollonius. These three mathematicians lived and worked within a span of only about one hundred years, Euclid about 300 B.C., Archimedes from about 287 to 212 B.C., and Apollonius about 200 B.C. This was well after the political influence of Greece had begun to wane (Alexander the Great died in 323 B.C.) and even longer after the culmination of Greek literature and art. It was characteristic for the times that none of the three lived in Greece proper. Euclid and Apollonius worked in Alexandria, Apollonius having been born in Perga in Asia Minor, while Archimedes lived in the Greek colony Syracuse on Sicily and was killed in 212 B.C. during the sack of this city by the Romans.

Greek mathematics thus reached its height in Hellenistic times, as the era following Alexander's death is called; but its beginnings are found some three centuries earlier. One of the most important and difficult problems for historians of Greek mathematics is the reconstruction of what happened before Euclid; for, with one rather trivial exception—a small work mostly on astronomy by one Autolycos—no complete mathematical text has survived from that period.

Euclid's *Elements* is thus the oldest Greek treatise on mathematics to reach us in its entirety, and the nature of this impressive work explains clearly why that is so. Euclid succeeded in incorporating practically all the mathematical knowledge accumulated by his predecessors into this one work, well arranged and well presented. The *Elements* therefore reduced the writings of the earlier mathematicians to matters of mere historical interest, and so they were not copied and hence are lost to us. It is characteristic that another work of Euclid, on conic sections, suffered a similar fate; for it was in turn superseded by Apollonius' brilliant treatise on *Conic Sections*, and all that remains of Euclid's contribution to this subject is its title.

During the last hundred years or so a number of scholars have devoted themselves to this difficult task of reconstructing pre-euclidean Greek mathematics. It involves a careful sifting of the classical Greek literature for references to early mathematicians and their works; if lucky, one may even find quotations from them. The knowledge gained in this fashion can be brought to bear on Euclid's *Elements*, and one may then venture to assign its components to various of Euclid's precursors. This task is far from finished, and occasionally an unexpected discovery, like that of Babylonian mathematics, forces upon us a re-evaluation of wide areas which, up to the time of the discovery, had seemed settled.

We shall not attempt to give a detailed account of early Greek mathematics here; so what follows is just the faintest outline of the pre-euclidean achievements.

According to Greek tradition (Herodotus and others) it was Thales of Miletus who in the beginning of the sixth century B.C. brought mathematics to Greece from Egypt, and who began to give mathematics the form it has retained since Greek antiquity by emphasizing the central position of the notion of proof. It is clear that many of the claims made for Thales are grossly exaggerated; for instance, Herodotus maintains that Thales predicted the fall of meteorites and a solar eclipse. Also, several of the proofs he was said to have devised are of such a nature that the accounts surely reflect the tastes of later times rather than the facts. But even allowing for the Greek penchant for hero-worship, there is little reason to doubt that Greek interest in mathematics began at the time of Thales, although we cannot say precisely what his achievements were. With what we now know of Egyptian and Babylonian mathematics, however, it seems likely that the initial impulse came from Mesopotamia rather than from Egypt.

There followed a century and a half of excited mathematical activity; we hear particularly about Pythagoras of Samos, who is supposed to have been in his prime about 530 B.C., and about his followers, the Pythagoreans. Their achievements were in science, particularly mathematics, and in religion; and their religious tenets were strongly flavoured by mathematical or number-mystical ingredients. Their tastes in mathematics tend toward arithmetic and algebra, and a strong Babylonian influence is obvious now that we know Babylonian mathematics. Indeed, Pythagoras is said to have visited Egypt and Babylon, and even though legend has it that he learned his mathematics in Egypt and his mystical beliefs in Babylonia, it is clear that it was from Babylonia that he got his mathematical inspiration.

Several other philosophers outside the ranks of the Pythagoreans were engaged in mathematical investigations; we know some of their names, such as Hippocrates of Chios and Democritus (both of the second half of the fifth century B.C.). A great many mathematical discoveries were made in algebra as well as in geometry. A few of these are discussed below.

Hippocrates' Lunes. In Simplicius we find an excerpt from Eudemus (fourth century B.C.), the author of a history of mathematics which unfortunately is lost. This excerpt is concerned with the so-called *lunes* (i.e. crescents) of Hippocrates, and below is recounted one of the three parts of this passage.

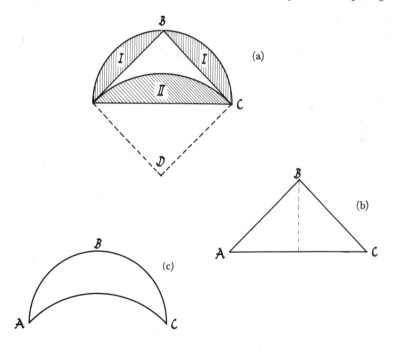

Figure 2.1

A semicircle is drawn on the diagonal AC of a square $ABCD$ (Figure 2.1a) and, with D as a center and AD as a radius, a 90° circular arc from A to C is drawn. Either of the segments I is a 90° segment of a circle, and so is segment II. They are therefore similar. Similar figures, however, have areas whose ratio is the *square* of their linear ratio. Thus

$$\frac{\text{segment I}}{\text{segment II}} = \left(\frac{AB}{AC}\right)^2 = \frac{AB^2}{AC^2}.$$

But this last ratio is 2, since AC is the diagonal of a square of side AB. Hence segment II is twice segment I, or is equal to the sum of the two segments I.

If we remove either the two segments I or segment II from the semicircle, we must end up with the same area; for, in either case we have removed the same amount. But in the first instance we obtain the triangle ABC (Figure 2.1b) and in the second the crescent or lune ABC (Figure 2.1c). The triangle and the lune must therefore have the same area. So we are able to *square* the lune. (To *square* a plane figure means to find a square of area equal to that of the given figure. Any polygon can readily be squared; in particular, to square the triangle ABC, we cut it along the altitude from B and rearrange the two new triangles to form a square.)

Hippocrates gives two other examples of squarable lunes; in one the outer arc is less and in the other greater than a semicircle.

These curious problems doubtless grew out of attempts at squaring the circle—we know now that this cannot be achieved with the severely restricted operations which we usually allow ourselves to perform with a pair of compasses and a straightedge—and it is indeed hinted that Hippocrates drew the conclusion from these examples that the circle also can be so squared. This is, of course, false; indeed, his lunes are expressly made so that they are squarable. But he succeeded in showing, for the first time, that there are domains with *curved* boundaries which are squarable, thus pointing out that the difficulty in squaring the circle is not merely that its circumference is not made up of straight line-segments.

Towards the end of the fifth century B.C. a critical reaction set in. It probably had a double origin: on the one hand in the discovery, most likely by the Pythagoreans, of the irrationality of what we call $\sqrt{2}$, and on the other, in the logical investigations begun by Parmenides and given sharp expression by Zeno in his famous paradoxes.

Irrationality of $\sqrt{2}$. The classical proof of the irrationality of $\sqrt{2}$ (a sketch of it appears as early as Aristotle) is as follows: The task is to show that *there is no fraction a/b, where a and b are integers, whose square is 2.*

We shall make use of the simple fact that the square of an even number is even, and the square of an odd number odd. The proof is indirect; that is, we pretend that the theorem is false and proceed to show that this assumption leads to a contradiction. Suppose then that there is a fraction whose square is 2; if there is such a fraction there is also such a reduced fraction. Let it be a/b. We then have

$$\frac{a^2}{b^2} = 2$$

or

(1) $$a^2 = 2b^2.$$

Now a must be an even number, for its square is even (namely $2b^2$) and the square of an odd number is odd. Since a/b is reduced, b must be odd; otherwise numerator and denominator would both be divisible by 2.

We found that a is even and b odd. This means that a is twice an integer, or

$$a = 2p, \qquad p \text{ integral,}$$

and substituting this in (1) we get

$$4p^2 = 2b^2$$

or

$$b^2 = 2p^2.$$

This is a contradiction; for, since b is odd, b^2 is also odd; but the last relation says that b^2 is even. Therefore our assumption must be false and our theorem true.

It is worthy of note that while the Babylonians found excellent sexagesimal approximations of $\sqrt{2}$ (see page 25) and presumably were content with these, the Greeks press the matter to its bitter logical end even though their result, that $\sqrt{2}$ is irrational, is of no great practical interest.

Zeno's Paradox about Achilles and the Tortoise. In his *Physics* Aristotle presents Zeno's paradoxes. The second is the so-called *Achilles* which in substance is the following:

The swift-footed Achilles is to run a race against a tortoise and, being fair, gives the tortoise a headstart. But contrary to his expectation (and our experience) Achilles is unable to overtake the tortoise. For, argues Zeno, when Achilles reaches the point where the tortoise was when he started, he has not yet overtaken the tortoise; for slow though it is, yet it has moved to a point farther on. When Achilles reaches that point, he still has not caught up with the tortoise; it is now at a point still farther on, and when he reaches that point, etc., etc. So he will never catch up with the creature.

Here Zeno has, by a clever device, divided the time interval from Achilles' start to his overtaking the tortoise into infinitely many time intervals, and then he maintains that a sum of infinitely many terms must needs be infinite, which is, of course, a fallacy. We might as well argue that $\frac{1}{3}$ is infinite, for

$$\tfrac{1}{3} = 0.333\cdots = 0.3 + 0.03 + 0.003 + \cdots,$$

where there are infinitely many terms on the right side.

Zeno's other paradoxes use the same sort of reasoning (e.g. the argument that motion is impossible, for, if an arrow does not move in an instant, then it cannot move in a time interval), and they are all related to problems from the domain of mathematics which today contains subjects such as continuity, limit processes, and a proper introduction of the real number system. Zeno's primary aim was probably to defend his, or rather Parmenides', philosophic system by showing how much more easily one could draw ridiculous conclusions from the tenets of rival systems, but his reasoning also holds up a warning example to mathematicians inasmuch as it shows how carefully a limit argument should be scrutinized before it is deemed convincing.

The discovery of the irrationality of $\sqrt{2}$ also belongs to the mathematical domain concerned with a proper understanding of real numbers and struck perhaps a more direct blow at mathematics. First it jeopardized the entire theory of similarity. This will become clear if, for a moment, we consider the fundamental theorem on similar triangles: *If corresponding angles of two triangles are equal, then corresponding sides are proportional.* Figure 2.2 should remind the reader of its elementary (and, as we shall see, incomplete) proof.†

We have two triangles, $\triangle ABC$ and $\triangle AB'C'$, in which

$$\angle A = \angle A$$

$$\angle B = \angle B'$$

$$\angle C = \angle C'.$$

Triangle $AB'C'$ has been placed so that its angle A coincides with angle A of triangle ABC and so that AB' lies along AB. $B'C'$ is then parallel to BC. Now let m be a line segment which goes p times into AB' and q times into AB, where p and q are integers ($p = 4$ and $q = 7$ in the figure). We then have

$$\frac{AB'}{AB} = \frac{p}{q}$$

† For related discussions see also I. Niven, *Numbers: Rational and Irrational*, Section 3.7, pp. 46–51, and L. Zippin, *Uses of Infinity*, Section 6.2, pp. 97–101; these are NML volumes 1 and 7, respectively.

and wish to show that then we must also have

$$\frac{AC'}{AC} = \frac{p}{q},$$

so that

$$\frac{AB'}{AB} = \frac{AC'}{AC}.$$

Drawing a series of parallels to $B'C'$ and BC (see Figure 2.2), and considering first the parallelograms and then the small triangles to the right, we can show that these triangles are congruent so that AC' is divided into p equal segments n and AC into q segments n. Hence

$$\frac{AC'}{AC} = \frac{p}{q}.$$

This proof is, of course, valid as far as it goes, but it rests on the assumption that one pair of corresponding sides, here AB' and AB, have a common measure m (or are *commensurable*). This means precisely that they have a *rational* ratio. But the proof of the irrationality of $\sqrt{2}$ showed clearly that this is not always so. For example, if AB' is the side of a square and AB its diagonal, the irrationality of $\sqrt{2}$ shows the impossibility of finding such an m. Thus the above proof does not apply to this case and is therefore incomplete.

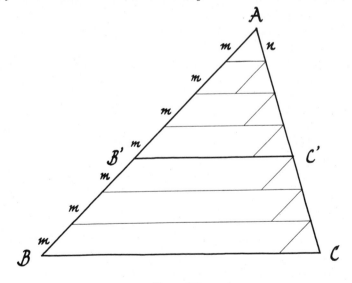

Figure 2.2

Secondly, the irrationality of $\sqrt{2}$ had serious consequences for algebra, for it showed that the simple problem of finding x such that

$$x^2 \;=\; 2,$$

which could easily be stated, had no exact solution in "numbers", for "numbers" then meant "rational numbers".

In the light of this new criticism with its implications for limit processes, similarity, and algebra, many of the old proofs lost their compelling force and were reduced to mere plausibility arguments. If mathematics was to be saved it had to be provided with a new and firmer foundation.

The algebraic dilemma probably gave rise to what we now call the *geometric algebra* of the Greeks. It will be noted that even though the equation

$$x^2 \;=\; 2$$

has no solution in rational numbers, it has a simple geometric solution: x is the diagonal in a unit square, as the Pythagorean theorem shows. The entire algebra was therefore reformulated in geometric terms; for instance, the phrase "the rectangle of sides a and b" was used instead of "a times b", and we are today adhering to this tradition when we call x^2 and x^3 the *square* and the *cube* on x. Book II of Euclid's *Elements* consists of theorems which on the surface belong to geometry, but whose contents are entirely algebraic. We shall see examples of this in the following presentation of the theorems leading up to Euclid's construction of the regular pentagon.

This discussion will also show that it is probable that one attempt at overcoming the difficulties about the notion of similarity consisted in avoiding it. But this could, of course, be only a temporary measure. It fell upon Eudoxos to provide a sound foundation for mathematics. We learn this from remarks by Archimedes and from Proclus' commentary to Euclid's *Elements*.

Eudoxos was a generation younger than Plato, so he must have flourished about 370 B.C. He rescued the theory of similarity by giving a new definition of *equality* of ratios and introducing a new criterion for when one ratio is *larger* than another. This he did entirely on the basis of whole numbers but in such a fashion that both rational and irrational ratios were covered. The ratios themselves he left undefined and this was very shrewd of him, for we know now that to define them would have involved a proper introduction of the real numbers. We find these things in Book V of Euclid's *Elements*.

By related methods Eudoxos devised a sound criterion for convergence of an infinite sequence (*Elements*, Book X, Proposition 1). It states:

Let two unequal magnitudes (e.g. lengths, or areas, or volumes) be given. From the larger, remove at least half; from the remaining part, remove at least half of it; and so on. After a finite number of steps, what remains will be less than the smaller of the two given quantities. Translated into modern symbolism, this proposition states:

Given $A > \varepsilon$, then if $\alpha_i \leq \frac{1}{2}$ $(i = 1, 2, \cdots)$, there exists an n such that

$$A \cdot \alpha_1 \cdot \alpha_2 \cdots \alpha_n \; < \; \varepsilon \;;$$

or, equivalently,

$$\text{if} \quad \alpha_i \leq \tfrac{1}{2}, \quad \text{then} \quad \lim_{n \to \infty} \alpha_1 \cdot \alpha_2 \cdots \alpha_n = 0.$$

Eudoxos' criterion was to form the sole foundation for all limit processes in the subsequent Greek mathematical works of a serious and scientific nature. In particular, it was the basis of what is called the method of exhaustion—a technique related to modern integration.

It is entirely outside the scope of this book to analyse Eudoxos' mathematical achievements, for to understand them properly takes a great deal of mathematical preparation and sophistication. It was not until the middle of the nineteenth century that a level of mathematical insight comparable to that of Eudoxos was reached once more.

In fact, the middle of the nineteenth century marks the end of a phase of mathematics which resembles the development outlined above. It began with the simultaneous invention by Newton and Leibniz of differential and integral calculus shortly before 1700. There then followed a century of excited activities by mathematicians who in their eagerness to claim new areas for mathematics could not be bothered by the niceties of the rules of rigorous procedure. Euler is the outstanding representative of this period. With incredible fertility he spouted forth masses of mathematical papers of the highest originality, but it was often only his unique intuition and insight that kept him from going wrong. But at the beginning of the nineteenth century a critical reaction set in, expressed by Abel and Cauchy among others, and during the latter half of the century the final steps towards solidifying the foundations of calculus were taken when Dedekind, Weierstrass, and Cantor introduced the real numbers in an unexceptionable fashion.

In mathematics, whether in the large or in the small, this is not an uncommon pattern: there is first a swift, sometimes inspired, and uncritical development, then a critical and doubting stage which necessitates meticulous work on the foundations, and finally a careful arrangement and polishing of the various parts giving the work its final form.

Euclid's great achievement in the *Elements* represents this last phase of early Greek mathematics, and his effort was so successful that the *Elements* remained a standard work for over 2000 years.

2.3   Euclid's Elements

We know nothing with certainty about Euclid except his preserved works. Even Proclus (410–485 A.D.), who wrote commentaries to the *Elements*, had to give plausibility arguments for assigning him to the reign of Ptolemy I Soter of Egypt (304–285 B.C.). He says that Euclid preceded Archimedes (287–212 B.C.), for Archimedes quotes the *Elements*, and that Euclid succeeded Eudoxos and Theaetetus, for their works are incorporated in the *Elements*. Since there is a story connecting Euclid with a King Ptolemy, Proclus concludes that it must be Ptolemy I.

The story is that the king, having looked through the *Elements*, hopefully asked Euclid if there were not a shorter way to geometry, to which Euclid severely answered: "In geometry there is no royal road!" This same story is, by the way, also told by Stobaeus about Alexander the Great and the mathematician Menaechmus; but then, it is a nice story. Stobaeus tells of Euclid that a student who had begun studying geometry with him asked, when he got through the first theorem, "What shall I gain by learning this?" whereupon Euclid called a slave and said, "Give him three-pence since he must needs make a gain of what he learns."

These anecdotes of dubious authenticity are all we know of Euclid's personality. In addition we can only say, as we saw, that he flourished about 300 B.C. in Alexandria and—but this is, of course, the only thing that really matters—that he wrote the *Elements*.

The *Elements* consists of thirteen Books, as they are called, and a translation of the text alone, without commentaries, would fill a large printed volume. In these thirteen books Euclid incorporates all the mathematical knowledge amassed at his time, with some notable exceptions such as conic sections and spherical geometry, and possibly some discoveries of his own. His great achievement is his presentation of material in a beautifully systematic form and his treatment of it as an organic whole.

He begins Book I with a list of *definitions*, of which the first is: "A point is that which has no part"; their purpose is to give the reader a feeling for the way mathematical terms will be used. There then follow five *postulates* and five *common notions*; together these form the assumptions on which the theory rests.

The postulates and the common notions are, in the translation of Heath [8],

Postulates.

Let the following be postulated:

1. To draw a straight line from any point to any point.
2. To produce a finite straight line continuously in a straight line.
3. To describe a circle with any center and distance.
4. That all right angles are equal to one another.
5. That, if a straight line falling on two straight lines make the interior angles on the same side less than two right angles, the two straight lines, if produced indefinitely, meet on that side on which are the angles less than the two right angles.

Common Notions.

1. Things which are equal to the same thing are also equal to one another.
2. If equals be added to equals, the wholes are equal.
3. If equals be subtracted from equals, the remainders are equal.
4. Things which coincide with one another are equal to one another.
5. The whole is greater than the part.

The Postulates are the basic assumptions peculiar to the specific branch of knowledge, in this case plane geometry, while the Common Notions are assumed in all fields. Today most mathematicians no longer see the need for such a distinction but call both kinds of assumptions *axioms* or *postulates*.

Before we discuss Euclid's axioms it is well to think a while of what it is that goes on in a body of mathematical theory. If we cut away the small talk of a mathematical treatise there remains a sequence of theorems, each followed by its proof. The proof consists in showing that the theorem in question is a logical consequence of previous theorems. But this means, of course, that the first "theorem" we put down cannot be proved, for there are no preceding theorems to be used in its proof. The unprovable theorems which initiate a theory are called *axioms*. It should be observed that it is irrelevant to the theory whether the axioms are "really" true or false; all the theory says is that *if* the axioms are true *then* so are all the subsequent theorems. Moreover, from the point of view of pure mathematics it makes no difference *what* the objects are that enter into the axioms, such as points, lines, circles; what matter are the *relations* between such undefined objects, e.g. that two distinct lines have at most one point in common. Far from reducing mathematics to mere idle talk this desistance from defining the objects with which a theory is concerned often makes it most useful. For, whenever objects satisfying the relations asserted in the axioms are encountered, for example in physics, the entire theory can be taken over and applied to these objects. This should not be taken to imply that any old set of axioms can set mathematicians working happily away. Apart from the demand that they should be "meaningful"—an ill-defined but most important facet—a set of axioms should possess the following three properties:

1. *Completeness*—this means that everything that will be used in the theory is fairly set down in the axioms so that there are no tacit assumptions.
2. *Consistency*—this means that it is impossible to derive two contradictory theorems from the axioms.
3. *Independence*—this means that no one of the axioms is a consequence of the others.

A few comments on each point might be in order. It has been mentioned above that the axiomatization of a theory often happens only after the theory has already been worked on for a while—this is a good way of assuring that the axioms are meaningful—and therefore it is not as simple as one might think to make sure that the axioms are complete; for, when one has grown accustomed to certain concepts it is beguilingly easy to forget that they must be justified by axioms. But if one proceeds with mincing steps, scrutinizing every argument minutely, one can ascertain that no tacit assumptions are made.

The consistency of a set of axioms is a most difficult thing to settle. One usually proves that a certain set of axioms, say that of plane Euclidean geometry, is consistent if another set of axioms is, say that for the real number system; in the given example the bridge between the two sets of axioms is analytic geometry. How far one can say that any one set of axioms is consistent is one of the questions which occupy mathematical logicians.

The last demand, that of independence, expresses a concern for economy, for it says that we should not put down any more axioms than we absolutely need. If a system of axioms is consistent and complete but not independent, no great harm is done; it merely means that one or more of the axioms are mislabeled and should rather be called theorems since they can be proved from the others.

These few remarks about axiomatics admittedly reflect a modern attitude, and although I have a feeling that some of the Greek mathematicians, in particular Archimedes, were more modern in their views than they are usually credited with having been, it is certain that, to most of the ancients, axioms appeared in a different light, as statements of facts and self-evident truths that everyone could subscribe to. Yet it is not improper to ask how Euclid's axioms measure up to the modern demands, for modern axiomatics is a descendant in direct line of Euclid's endeavors.

Already the first proposition in the first book shows clearly that Euclid's axioms are not complete. The proposition is to construct an equilateral triangle on a given straight line-segment *AB*. The construction is the usual one (see Figure 2.3). There is, however, nothing in the axioms that entitles us to conclude that the two circular arcs have a point in common, so Euclid has made a tacit assumption. If we say that *obviously* the circles must intersect or that any one can try for himself and *see* that they do so, we are no longer doing mathematics but are rather conducting an empirical study of graphics (where, by the way, the axioms do not hold). Diagrams are, of course, useful; they can give us many ideas and may help us hold on to the thread of an argument, but they must never be a link in our reasoning.

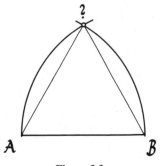

Figure 2.3

Euclid makes several other tacit assumptions, notably about displacement and congruence, but we have already mentioned how difficult it is to recognize problems when our keenness of vision has been dulled by habit. In fact, it was not until 1900 that a complete set of axioms for Euclidean geometry was given (by David Hilbert in his famous "Grundlagen der Geometrie," i.e. Foundations of Geometry)

Euclid's axioms are as consistent as arithmetic, for we can build an arithmetical model which satisfies them. This model is called analytic geometry.

The axiomatic question of greatest concern to mathematicians from antiquity until the middle of the nineteenth century was the independence of Euclid's system, specifically of the fifth postulate. This seems somewhat curious for, as we saw, the independence of the axioms has no impact on the logical validity of the theory as a whole, but this activity reflects the attitude of the older mathematicians toward axiomatics.

The fifth postulate is called the *parallel postulate,* for it implies immediately that through a given point P outside a line l passes at most one line parallel to l; for, if (see Figure 2.4)

$$u + v \; < \; 2 \times 90°,$$

the fifth postulate states that l' and l will intersect, and if

$$u + v \; > \; 2 \times 90°,$$

it is easily seen that l' and l will intersect on the other side. Thus the only line through P which has a chance of being parallel to l is l'' for which

$$u + v \; = \; 2 \times 90°.$$

That l'' is, in fact, parallel to l follows from the other axioms (Euclid proves this in I, 28†).

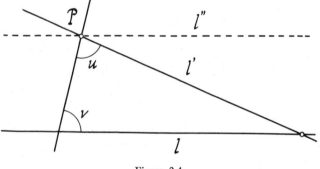

Figure 2.4

It was somehow felt that the fifth postulate was not as natural or self-evident as the others, for it spoke of the existence of a point of intersection which might well be thousands of miles away. And so numerous attempts at proving it were made, e.g., one by Ptolemy, the astronomer. But a close scrutiny of these "proofs" of the parallel postulate reveals that their authors merely succeeded in replacing it by various tacit assumptions which to them seemed less offensive.

Towards the end of the eighteenth century, beginning with Saccheri's work, new attempts were made at showing the dependence of the parallel postulate, now by means of indirect proofs. One reasoned thus: if the

† Here and throughout, in references to the *Elements,* the Roman numeral refers to the Book, the Arabic numeral to the Proposition in question.

parallel postulate is a consequence of the first four postulates, then these first four and the *negation* of the parallel postulate will lead to a contradiction, that is, they will be inconsistent. But far from leading to a contradiction, this new set of axioms turned out to be the basis of a beautiful and consistent theory for which arithmetical models may be built—the beginning of what is now called non-euclidean geometry. It was thus shown that the first four postulates are compatible with both the parallel postulate and its negation, and therefore the independence of Euclid's postulates was at last established, more than two thousand years after his death. The mathematicians who founded non-euclidean geometry were, in the first rank, Gauss, Bolyai, and Lobachevsky.

Euclid was thus vindicated by non-euclidean geometry. He had been right in making the parallel postulate one of his axioms. That this decision had not come to him lightly can be inferred from the way he used it—we must here, as always with Euclid, make inferences from the text, for the *Elements* contains no prefaces and no commentaries or justifications, but only definitions, axioms, theorems, and proofs. In Book I he does not employ the parallel postulate until he reaches the proof of Proposition 29, which states that when two parallel lines are cut by a third then the interior angles on the same side are equal to two right angles (the proof is essentially the argument given above). He had had an inviting chance to use the fifth postulate immediately after I, 17, and if he had done so he could have shortened and even sharpened several of the following arguments. It is thus clear that the delay is deliberate and that Euclid chose to prove as much as he possibly could without using the parallel postulate though this meant slower progress. We see here on the one hand evidence of Euclid's special feelings about the parallel postulate, and on the other our first example of his devotion to the principle of economy of means. We can also note that Euclid, in the first 29 propositions of Book I, is doing euclidean and one kind of non-euclidean geometry at the same time.

A summary of the contents of the thirteen books of the *Elements* is as follows:

Book I	Elementary constructions, congruence theorems, area of polygons, Pythagorean theorem
Book II	Geometric algebra
Book III	Geometry of the circle
Book IV	Construction of certain regular polygons
Book V	Eudoxos' theory of proportions
Book VI	Similar figures
Books VII–IX	Number theory

An old-fashioned geometry course would cover most of Book I, a few selections from Books III and IV, and a diluted presentation of some of the theorems in Book VI. A glance at the above list of contents will give some idea of the scope of Euclid's work, and the presence of Books II, VII, VIII, IX, and X shows clearly that the identification of the words "Euclid" and "geometry", which used to be made in English, is not entirely justified. Books II and X are algebraic in substance, and Books VII–IX deal with number theory, i.e. the branch of mathematics which is devoted to the study of whole numbers, or integers. We have seen evidence of Babylonian interest in number theory (Plimpton 322 and Pythagorean numbers; see Chapter 1, p. 30), but, as everywhere, it is not until Euclid that we find a logical sequence of general theorems with proper proofs. It is characteristic of integers (in contrast to rational numbers), that one number does not always go into, or divide, another. Euclid is particularly interested in divisibility theory, and he properly emphasizes the role played in it by *prime numbers*, or simply *primes*.†
We find, among many other things, a method for determining the greatest common divisor of two integers (now called *Euclid's algorithm*) and a proof of the theorem that there are infinitely many primes, i.e. that the sequence 2, 3, 5, 7, 11, 13, 17, $\cdots$ never ends, or, in Euclid's words (IX, 20):

Prime numbers are more than any assigned multitude of prime numbers.

The proof is so classically beautiful that I must give it here, in substance. It presupposes

VII, 31: *Any composite number is measured by a prime*, i.e., a composite number (a number $a \neq 1$ which is not a prime) has always a prime factor. For, that a number a is composite means that it has a factor d less than a and greater than 1. If d is a prime our theorem is proved, and if not, d is composite and has a factor d' less than d and greater

† A prime number is an integer greater than 1 that is divisible only by 1 (and itself), i.e., an element of the sequence 2, 3, 5, 7, 11, 13, $\cdots$. The name is a literal translation of the Greek term *protos arithmos*. There are technical reasons for not counting 1 among the primes.

than 1. d' is then also a factor in a. If d' is a prime the proof is complete, and if not, d' has a factor d'', less than d' and greater than 1. d'' is a factor of d', hence of d, and hence of a. If d'' is a prime the theorem is proved, etc. But this cannot go on indefinitely, for

$$a > d > d' > d'' > \cdots$$

is a decreasing sequence of whole numbers, all greater than 1, and it must therefore be finite (it can certainly have no more than a members). The only way in which it can end is by some d being a prime, so a has a prime factor.

We now turn to Euclid's proof of the infinitude of primes. He shows that if we are given certain primes $p_1, p_2, \cdots, p_n$ we can always find one more. For let us consider the number N obtained by multiplying together the given primes and then adding 1, i.e.

$$N = p_1 \cdot p_2 \cdots \cdot p_n + 1.$$

If N happens to be a prime, we have found a new prime, for N is larger than any of the primes $p_1, p_2, \cdots, p_n$. But even if N is not a prime we can draw the desired conclusion; for, if N is composite it has, as just shown, a prime factor, say p. This prime factor p cannot be p_1, because division of N by p_1 leaves the remainder 1, nor can p be p_2, nor any other of the given primes for the same reason, namely that division by any of the given primes leaves the remainder 1. So p is a new prime.

Thus, whenever a certain finite set of primes is given, we can always be sure that there is one more, and hence that there are infinitely many primes.

These two arguments are Euclid's, although the words and modern notation are not. I shall follow the same practice of giving Euclid's arguments in somewhat modern guise in the following proofs. I have selected the sequence of theorems in the *Elements* which lead up to the construction of a regular pentagon, omitting the more trivial ones. The reader who wishes to see the form and style of Euclid's own presentation is referred to Heath's excellent English edition of the *Elements* [8].

The construction of a regular pentagon has, of course, considerable interest in itself, but Euclid is also going to use it in Book XIII when he wishes to construct a regular dodecahedron (a solid bounded by twelve congruent regular pentagons). Further, the pentagon construction plays an important role in the computation of entries in Greek trigonometric tables, as we shall see in Chapter 4.

2.4 Euclid's Construction of the Regular Pentagon

Before we turn to Euclid's construction of the regular pentagon, it is well to take a look at the way this problem is commonly treated today. It is clear that once we succeed in constructing an angle of 72° we have solved our problem, for $72° = 360°/5$ is the central angle in the pentagon.

Actually, it so happens that it is easier to attack the problem of constructing a regular decagon (polygon of ten sides), that is, to construct an angle of $36° = 72°/2$, because of the singularly nice properties of the central triangle OAB in Figure 2.5.

We wish, then, to inscribe a regular decagon in a given circle of centre O and radius r. We shall try to express the desired side length x of the decagon in terms of the given length r in such a way that x can be constructed with straightedge and compasses. The isosceles triangle OAB in Figure 2.5 has as its base AB a side x of the decagon, as its sides $OA = OB$ two radii r of the circle, and as its vertex angle O the central angle of a regular decagon (which is 36°).

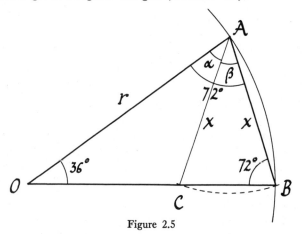

Figure 2.5

We first observe that each of the two equal angles OAB and OBA is 72°, for the sum of the angles in a triangle is 180°. We now find a point C on OB such that $AC = x$ by drawing an arc with A as centre and AB as radius. Thus $\triangle ABC$ is isosceles, so that

$$\angle ACB = \angle ABC = 72°.$$

This leaves 36° for β, so that $\alpha = 72° - 36° = 36°$.

Since now $\triangle CAO$ has equal angles (of 36°) at A and O, it is isosceles, and we conclude that $OC = x$. Thus $CB = r - x$.

We now note that our original triangle OAB is similar to $\triangle ABC$; and we obtain

(1)
$$\frac{r}{x} = \frac{x}{r-x},$$

whence

$$x^2 + rx - r^2 = 0.$$

Solving this quadratic equation for x, we get

$$x = \tfrac{1}{2}(-r \pm \sqrt{r^2 + 4r^2}),$$

and after discarding the negative solution,

(2)
$$x = \tfrac{1}{2}r(\sqrt{5} - 1).$$

Our analysis leads to an expression for the side of the decagon in terms of the radius r of the circumscribed circle; if r is given, x can be computed.

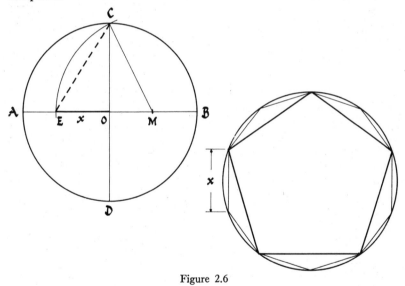

Figure 2.6

We shall now give a construction for x by straightedge and compasses which reproduces the algebraic expression (2). Let AB and CD be two perpendicular diameters in a circle of centre O and radius r; see Figure 2.6. Furthermore, let M be the midpoint of OB. With M as a centre and MC as a radius we draw a circle which intersects OA at E. We shall show that OE is the side of the regular decagon inscribed in our circle.

It follows from the Pythagorean theorem that

$$MC \;=\; \sqrt{OM^2 + OC^2} \;=\; \sqrt{(\tfrac{1}{2}r)^2 + r^2} \;=\; \tfrac{1}{2}r\sqrt{5};$$

hence

$$OE \;=\; ME - MO \;=\; MC - MO \;=\; \tfrac{1}{2}r(\sqrt{5} - 1)$$

which is indeed in agreement with (2).

Thus OE will go ten times as a chord into the circle of radius r, and the regular pentagon can easily be drawn if we select every other point as a vertex.

Problem

2.1 Show that EC in Figure 2.6 is the side of the regular pentagon inscribed in a circle of radius OA. (This is equivalent to Euclid XIII, 10, which says that the sides of the regular pentagon, hexagon, and decagon in a given circle form a right triangle.)

Let us now for a moment look back upon this proof to see what sort of mathematical tools we employed beyond the more elementary geometrical theorems such as the one warranting that the sum of the angles in a triangle is 180°. We observe that in order to establish (1) we used the principal theorem from the theory of similarity: *If corresponding angles of two triangles are equal, then their corresponding sides are proportional.* To get from (1) to (2), we made use of the formula giving the solution of a quadratic equation. And we may add that, in our verification of the construction, we employed the Pythagorean theorem.

When we now turn to Euclid's solution of this same problem, we shall pay particular attention to the tools he chose to use. It may seem curious to some to make so much fuss about what goes into a proof; after all, one may say that the important thing about a proof is that it is valid and yields the desired result. It is, of course, difficult to argue against such an attitude, for we are now entering the lawless realm of taste. Yet you will hear mathematicians talk about beautiful proofs and ugly proofs, elegant proofs and clumsy proofs; and aesthetic pleasure is not the least satisfaction a mathematician derives from his work.

It is, of course, impossible to reach agreement on what constitutes mathematical beauty and elegance, but some of the more common ingredients are such aspects as brevity, economy of means, surprising and dramatic turns, clarity, new applications of old techniques, and methods that lend themselves to generalization to other situations. Some of these ingredients are sometimes at odds, such as brevity and economy of means.

To prove a good theorem with the weakest possible tools is rather like landing a large trout on an old and beloved silk line. It does not make for speed or brevity, but it has an undeniable charm. Euclid is not always given to swiftness, but he is rather devoted to the task of getting as much as he can with as little as he can get away with.

The first of Euclid's theorems in the sequence leading to the pentagon construction and the one which holds the clue to his success as a miser of tools is the following:

THEOREM 1 (I, 35) *Parallelograms on the same base and between the same two parallel lines are equal (in area).*

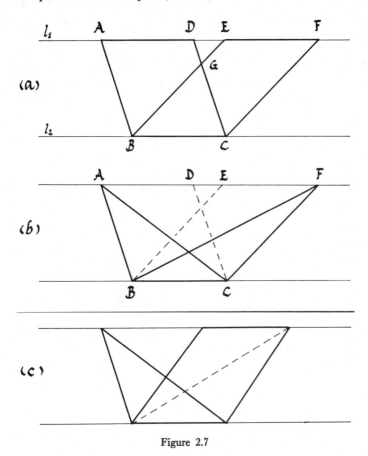

Figure 2.7

Let the two parallelograms be *ABCD* and *EBCF* on the common base *BC* and between the two parallel lines l_1 and l_2 (Figure 2.7a).

Euclid first shows that the two triangles ABE and DCF are congruent. He then obtains his desired result by observing that

$$\square\, ABCD \;=\; \triangle BCG + \triangle ABE - \triangle DGE$$

and

$$\square\, EBCF \;=\; \triangle BCG + \triangle DCF - \triangle DGE,$$

where, as we saw, the two middle terms on the right are equal, and so the parallelograms are equal.

I have never been able to understand why Euclid chose this proof instead of the following:

If from the entire figure $ABCF$ we subtract triangle DCF, we leave parallelogram $ABCD$; if from the same figure we subtract triangle ABE, we leave parallelogram $EBCF$. But in both cases we subtracted the same amount, so the parallelograms are equal.

This proof is not only shorter than Euclid's but it is even more general, for it does not presuppose that the lines BE and DC intersect between the parallels l_1 and l_2, which Euclid's does. I cannot help hoping that the text is corrupt and that the second version is, in fact, Euclid's.

THEOREM 2 (I, 37) *Triangles on the same base and between the same parallel lines are equal.*

This theorem follows immediately from Theorem 1. Let the triangles be ABC and FBC (Figure 2.7b). Triangle ABC is half of parallelogram $ABCD$, and triangle FBC is half of parallelogram $EBCF$. But the two parallelograms are equal; hence so are the triangles.

A similar, and equally simple argument proves

THEOREM 3 (I, 41) *If a parallelogram and a triangle have the same base and are between the same parallels, then the parallelogram is twice the triangle* (Figure 2.7c).

THEOREM 4 (I, 43) *In any parallelogram the complements of parallelograms about the diagonal are equal.*

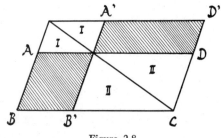

Figure 2.8

Parallelograms about the diagonal are such as the ones composed of the two triangles I and of the two triangles II, respectively, and their complements are the two shaded parallelograms (see Figure 2.8). These complements are equal, for either remains when we subtract a triangle I and a triangle II from half of the large parallelogram.

While Theorem 1 enables us to transform a parallelogram into another of the same area and the same base but with different angles, Theorem 4 makes it possible to transform a parallelogram into another of the same area and the same angles, but different sides; parallelograms $ABCD$ and $A'B'CD'$ of Figure 2.8 are so related.

EXAMPLE: Solve by construction the equation

$$x \cdot a = b \cdot c,$$

where a, b, c are given line segments.

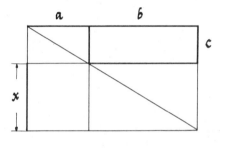

Figure 2.9

The solution is given in Figure 2.9, for the complements about the rectangle's diagonal are $x \cdot a$ and $b \cdot c$, respectively, and they are equal according to Theorem 4. (The reader should take pains to supply all the steps of the construction, beginning with the given quantities a, b, and c.)

We usually solve this problem by transforming the equation to

$$\frac{a}{b} = \frac{c}{x},$$

and we then find x as the fourth proportional to $a, b,$ and c. We note, however, that this procedure involves proportions, a concept which Theorem 4 enabled us to avoid. Theorem 4 also made possible a geometric solution of an equation; we have here the first example of the Greek geometrical algebra. This problem and its solution is a special case of Euclid's I, 44.

THEOREM 5 (I, 47, Pythagoras' theorem) *In right-angled triangles the square on the side subtending the right angle (i.e. the hypotenuse†) is equal to (the sum of) the squares on the sides containing the right angle.*

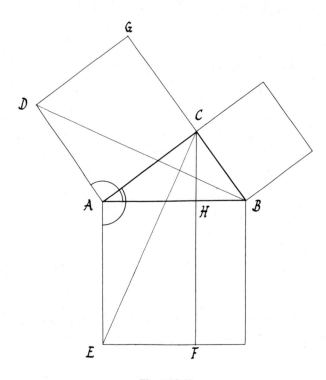

Figure 2.10

Euclid's proof is as follows: On the three sides of a right triangle *ABC* (∡*C* = 90°) squares are constructed (see Figure 2.10). The altitude *CH* from *C* is drawn and continued to *F*. Segments *DB* and *CE* are drawn.

First observe that triangle *DAB* is congruent to triangle *CAE*; to see this, note that if we turn one by 90° about *A* it will cover the other.

Now the square on *AC* is twice triangle *DAB*, for they have the same base (*AD*) and lie between parallel lines (Theorem 3); similarly we see that the rectangle *AEFH* is twice triangle *CAE*, for they too have the same base (*AE*) and lie between parallel lines. Since the two triangles are congruent, the square on *AC* is equal to the rectangle *AEFH*.

† Hypotenuse comes from the Greek *hypotenousēs* = subtending.

It follows in exactly the same manner that the square on BC is equal to the rectangle under HB, and so the sum of the *squares on AC* and BC is equal to the sum of the two rectangles; but that is exactly the square on AB.

This is a very elegant proof (despite Schopenhauer's foolish remarks†). First we note that, as before, the concept of similarity has been avoided; secondly that, in addition to the Pythagorean theorem, Euclid has proved that *in a right triangle the square of a side is equal to the product of its projection on the hypotenuse and the entire hypotenuse*. Once we have introduced proportions, this is equivalent to saying that *a side in a right triangle is the mean proportional between its projection on the hypotenuse and the entire hypotenuse.*

We now turn to two theorems from Book II. This book is concerned with what we now call geometrical algebra; what is meant by this will become apparent as we study the next theorem. For once, I shall quote a word-true translation (by Heath) of the awesome original:

THEOREM 6 (II, 6) *If a straight line be bisected and a straight line be added to it in a straight line, the rectangle contained by the whole with the added straight line and the added straight line together with the square on the half is equal to the square on the straight line made up of the half and the added straight line.*

To throw some light into the darkness of this theorem, consider Figure 2.11. The given straight line is AB, and it is bisected at D. The added straight line is BC. The theorem now says that the rectangle of sides AC and BC, i.e. $AC \cdot BC$, plus the square on DB (or AD) is equal to the square on DC, or, expressed with the small Greek letters,

$$(1) \qquad (2\alpha + \beta)\cdot\beta + \alpha^2 \;=\; (\alpha + \beta)^2.$$

The proof is simple enough if we use Figure 2.11. The left side of (1) consisting of the long rectangle and the square is

$$(\mathrm{I} + \mathrm{II} + \beta^2) + \alpha^2,$$

and the right side is

$$\mathrm{II} + \mathrm{III} + \beta^2 + \alpha^2.$$

---

† Schopenhauer calls it a "mouse-trap proof" and also "des Eukleides stelzbeiniger, ja, hinterlistiger Beweis"; freely translated, this means "Euclid's artificial, stilted, in fact sly and underhanded proof."

But I and III (the shaded areas) are equal and hence so are the two sides of (1). By the way, to prove that I = III Euclid uses that II and III are complements about the diagonal of the square on DC and therefore equal (by Theorem 4); I = II because they are congruent.

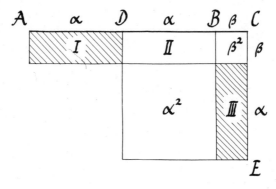

Figure 2.11

It should be clear, once we have translated the geometrical verbiage in Theorem 6 into the modern notation of (1), that this is not a theorem in geometry but rather an algebraic identity—and an important one— in a clumsy geometrical garb. But, as we saw above, there were good reasons for going to this length. We shall recognize the importance of Theorem 6 when we see how often it is employed in what follows.

The theorem is perhaps easier to remember if we set

$$DC = a \quad \text{and} \quad AD = DB = b,$$

so that

$$AC = a + b \quad \text{and} \quad BC = a - b;$$

for then it says

$$(a + b)(a - b) + b^2 = a^2$$

which is a slight variation on our well-known identity

$$(a + b)(a - b) = a^2 - b^2.$$

THEOREM 7 (II, 11) *To cut a given straight line so that the rectangle contained by the whole and one of the parts is equal to the square on the other.*

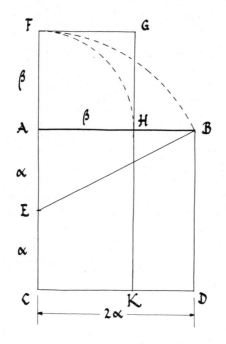

Figure 2.12

Let the given line be AB; see Figure 2.12. The problem is to find a point H on it so that

$$AB \cdot HB = AH^2.$$

The solution is as follows: On AB a square $ABDC$ is constructed and the side AC is bisected at E. AC is extended beyond A. EB is swung around E to F on the extension of AC. On AB, AH is marked off so that AH is equal to AF.

We now maintain that H is the desired point. To see this, we complete the square $AFGH$, and extend GH to K. We begin our argument by using Theorem 6 on AC, bisected at E, and with AF added on. It yields

$$FC \cdot FG + AE^2 = EF^2$$

or

$$(2\alpha + \beta) \cdot \beta + \alpha^2 = (\alpha + \beta)^2.$$

Observing that

$$EF = EB$$

and using Theorem 5 (Pythagorean theorem) we get

$$FC \cdot FG + AE^2 \ = \ EB^2 \ = \ AE^2 + AB^2$$

or

$$(2\alpha + \beta) \cdot \beta + \alpha^2 \ = \ \alpha^2 + (2\alpha)^2.$$

Subtracting AE^2 from both sides we get

$$FC \cdot FG \ = \ AB^2,$$

or

$$(2\alpha + \beta) \cdot \beta \ = \ (2\alpha)^2,$$

or again, in geometrical terms: the long rectangle contained by FC and FG is equal to the square on AB. Euclid now subtracts from each of these equal figures their common part which is the rectangle contained by AC and AH and leaves, on the one hand, the square on AH, and on the other, the rectangle contained by HB and BD; they must therefore be equal, so

$$AH^2 \ = \ HB \cdot BD.$$

But $BD = AB$, so

$$AH^2 \ = \ HB \cdot AB,$$

or

$$\beta^2 \ = \ (2\alpha - \beta) \cdot 2\alpha,$$

which was to be proved.

If we call $AB = a$, i.e. $2\alpha = a$, we see that β is a solution of the quadratic equation

$$x^2 \ = \ (a - x) \cdot a$$

or

$$x^2 + ax - a^2 \ = \ 0.$$

We recognize both this equation and its solution from the modern solution of the pentagon construction which introduced this section; see page 55.

So we see that in this theorem Euclid has actually solved a quadratic equation in a manner quite different from the methods of the Babylonians. But the solution

$$x \ = \ \tfrac{1}{2}a(\sqrt{5} - 1)$$

involves, as we know, an irrational number, so that the geometrical procedure was necessary when irrational numbers were not known.

Theorem 7 is a special case of theorems—appearing particularly in Book VI—wherein different forms of general quadratic equations are solved by means of theorems like 4 and 6. One recognizes the equations as well-known Babylonian types, but the geometrical procedure permits irrational solutions, although it excludes negative solutions.

After proportions have been introduced, Theorem 7 reappears as VI, 11 in connection with the problem of cutting a given straight line segment in *mean and extreme ratio*. This means to cut a line segment a so that one of the parts, x, is the mean proportional between a and the other part, $a - x$, i.e. so that

$$\frac{a}{x} = \frac{x}{a - x}$$

or

$$x^2 = a(a - x),$$

which is precisely our equation. The point H in Figure 2.12 thus divides AB in mean and extreme ratio. This cut is called the *golden section*, a term of fairly recent origin; in antiquity it was simply called *the* section. It was supposed to be particularly harmonious and pleasing to the eye.

We now turn to three theorems from Book III which is concerned with the geometry of the circle.

THEOREM 8 (III, 36) *If from a point P outside a circle we draw a line tangent to the circle at T and an arbitrary line intersecting the circle at R and S, then we shall always have*

$$PR \cdot PS = PT^2.$$

This theorem says that for a given circle and a given point P the product $PR \cdot PS$ is *constant* and equal to PT^2; this value is sometimes called the *power* of the point with respect to the circle.

In the proof Euclid considers two cases which together exhaust all possibilities:

1. the line from P passes through the center C, and
2. the line from P does not pass through C.

Case 1 is illustrated in Figure 2.13a. RS is bisected at C and PR is added on so that, by Theorem 6,

$$PR \cdot PS + RC^2 = PC^2.$$

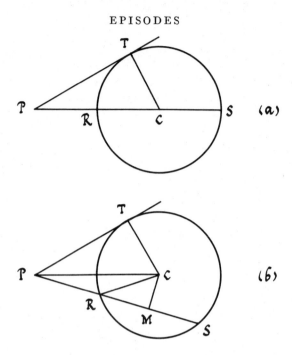

Figure 2.13

Now since RC and TC are radii of the circle,

$$RC^2 = TC^2,$$

and when this equation is subtracted from the preceding one the result is

$$PR \cdot PS = PC^2 - TC^2.$$

But by the Pythagorean theorem

$$PC^2 - TC^2 = PT^2,$$

whence the desired result

$$PR \cdot PS = PT^2.$$

For case 2 we consider Figure 2.13b. M is the midpoint of RS and CM is perpendicular to RS. Applying Theorem 6 once again one gets

$$PR \cdot PS + RM^2 = PM^2.$$

Adding CM^2 to both sides we obtain

$$PR \cdot PS + (RM^2 + CM^2) = (PM^2 + CM^2).$$

An application of the Pythagorean theorem to the quantities in the parentheses reduces this to

$$PR \cdot PS + RC^2 \;=\; PC^2$$

which was the starting point in case 1. So, by subtracting

$$RC^2 \;=\; TC^2,$$

the desired relation

$$PR \cdot PS \;=\; PC^2 - TC^2 \;=\; PT^2$$

is obtained.

We shall, however, rather make use of a converse of Theorem 8, which we now state and prove.

THEOREM 9 (III, 37) *If, from a point A outside a circle, two lines are drawn, one intersecting the circle at B and F and the other meeting the circle at D, and if*

$$AB \cdot AF \;=\; AD^2,$$

then AD is tangent to the circle at D.

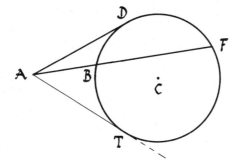

Figure 2.14

The proof is very simple. From A we draw a line AT, tangent to the circle at T; see Figure 2.14. From Theorem 8 we know that

$$AB \cdot AF \;=\; AT^2.$$

But this means that

$$AT \;=\; AD.$$

It follows from the symmetric position of AT and AD with respect to the line AC through the centre of the circle that D is a point of tangency.

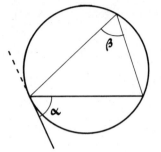

Figure 2.15

Figure 2.15 should serve to remind the reader of the following theorem.

THEOREM 10 (III, 32) *The angle α between a tangent and a chord of a circle is equal to the angle β contained in the arc determined by the chord and on the side of the chord opposite from α.*

Problem

2.2 Give a careful formulation and a proof of this theorem.

We are now, finally, prepared to come to grips with the pentagon construction.

THEOREM 11 (IV, 10) *To construct an isosceles triangle having each of the angles at the base twice the remaining one.*

Such a triangle would solve our problem, for its angles must be 36°, 72°, 72°, and 72° is a fifth of 360°.

Let there be given the straight line AB (Figure 2.16). First, according to Theorem 7, a point C can be constructed on AB so that

(1) $$AB \cdot CB = AC^2.$$

A circle is drawn with A as centre and AB as radius, and the chord BD is marked off so that

$$BD = AC;$$

The segments AD and CD are drawn. Triangle ABD is isosceles, so

(2) $$\beta = \gamma + \delta.$$

The next task is to show that triangle ABD has the desired property or that

$$\beta = \gamma + \delta = 2\alpha.$$

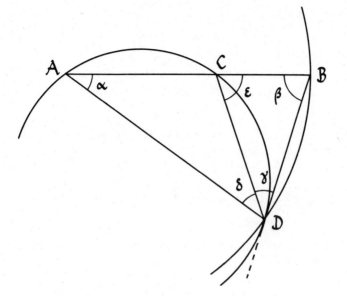

Figure 2.16

To that end we first construct triangle ACD's circumscribed circle, which we shall call the small circle. If we now substitute BD for AC in (1) we get

$$CB \cdot AB = BD^2.$$

This means, however, according to Theorem 9, that BD is tangent to the small circle at D. But then Theorem 10 tells us that

$$\alpha = \gamma.$$

All we need to show then is that

$$\alpha = \delta.$$

Since

$$\alpha = \gamma,$$

we have

(3) $$\alpha + \delta = \gamma + \delta.$$

But since ε is a supplementary angle to angle C in triangle ACD, it is equal to the sum of the other two angles (for this sum added to angle

C also gives 180°), i.e.

$$\varepsilon \;=\; \alpha + \delta$$

and so also, by (3)

$$\varepsilon \;=\; \gamma + \delta,$$

and by (2)

$$\varepsilon \;=\; \beta.$$

But this means that triangle CDB is isosceles, so CD equals BD which was made equal to AC. Thus we have

$$CD \;=\; CA$$

or that triangle ACD is isosceles. Hence

$$\alpha \;=\; \delta$$

which was all that remained to be proved. For we showed already that

$$\alpha \;=\; \gamma$$

and so the angles at the base BD of the isosceles triangle ABD are twice the angle A. Hence angle A is 36°, and the construction of the pentagon is then a simple matter which Euclid performs in IV, 11, and of which I include only the diagram (Figure 2.17).

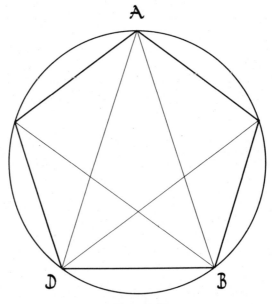

Figure 2.17

Euclid finally constructs the regular quindecagon (15-gon) in IV, 16; see Figure 2.18. To convince ourselves of the validity of the construction, we need only observe that

$$\frac{2}{5} \cdot 360° - \frac{1}{3} \cdot 360° = \frac{1}{15} \cdot 360°.$$

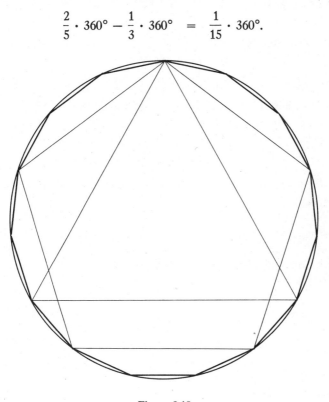

Figure 2.18

Let us look back upon this sample of Euclidean mathematics as a whole. It is obvious that it is different from Babylonian mathematics in aim as well as in method. Yet we recognize in this structure several Babylonian elements, among them the Pythagorean theorem and quadratic equations, but in new garb.

Euclid is not interested in drill methods for solving trivial variations on standard problems. He wants to establish general theorems, strong enough to enable him to solve crucial problems such as the pentagon construction, and he is greatly concerned with the structure of the chain of results connecting a given theorem to his axioms. We noticed in particular that he wanted this chain to contain as few notions as possible for he deliberately avoided the use of similarity arguments. Where we find it natural to set up a proportion

$$\frac{a}{b} = \frac{c}{d}$$

between line segments to obtain

$$a \cdot d = b \cdot c,$$

Euclid's device is to derive this second form directly. He can do so because of his powerful theorems on areas of polygons.

We saw further that he solved ensuing equations directly in geometrical form; he had to do this because he had no way of expressing the solution

$$x = \tfrac{1}{2}r(\sqrt{5} - 1)$$

which involves an irrational number and which we used as an intermediary step.

But above all we recognize that this is genuine mathematics. It can be enjoyed immediately without any allowances for its venerable age, and it is still part of the living organism of mathematics. A capacity for creating and delighting in such intellectual structures is one of the faculties that set man apart from the beasts.

CHAPTER THREE

Three Samples of Archimedean Mathematics

3.1 Archimedes' Life

In weightiness of matter and elegance of style, no classical mathematical treatise surpasses the works of Archimedes. This was recognized already in antiquity; thus Plutarch says of Archimedes' works:

> It is not possible to find in all geometry more difficult and intricate questions, or more simple and lucid explanations. Some ascribe this to his genius; while others think that incredible effort and toil produced these, to all appearances, easy and unlaboured results.

Plutarch, who lived in the second half of the first century A.D., writes this in his *Lives of the Noble Grecians and Romans*, more specifically in his life of Marcellus. Marcellus was the general in charge of the Roman army that besieged, and ultimately took, the Greek colony of Syracuse on Sicily during the second Punic War (218–201 B.C.). Archimedes' ingenious war-machines played an important role in the defense of Syracuse, and for this reason Plutarch writes about him at some length.

Archimedes introduces each of his books with a dedicatory preface, where he often gives some background for the problem he is about to treat. These prefaces contain precious information for the historian of mathematics, and they even throw some light on Archimedes' life. There are, furthermore, scattered references to him in the classical literature, and so he becomes the Greek mathematician about whom we have the most biographical information, even though it is precious little.

Archimedes was killed in 212 B.C. during the sack of Syracuse that ended the Roman siege. Since he is said to have reached the age of 75 years, he was born about 287 B.C. In the preface to his book *The Sand-reckoner*, he speaks of his father Pheidias, the astronomer, who is otherwise unknown. It is said that Archimedes studied in Alexandria, then the centre of learning, and it is certain that he had friends among the Alexandrian mathematicians, as we learn from his prefaces; but he spent most of his life in Syracuse where he was a friend and, as some even say, a relation of the reigning house.

He spent his life pursuing interests which extended from pure mathematics and astronomy to mechanics and engineering. Indeed, it was his more practical achievements that caught the public fancy. If we may trust the stories about him, he was not above adding a dramatic touch to his demonstrations; thus Plutarch tells, in Dryden's stately translation:

> Archimedes, however, in writing to King Hiero, whose friend and near relation he was, had stated that given the force, any given weight might be moved, and even boasted, we are told, relying on the strength of demonstration, that if there were another earth, by going into it he could remove this. Hiero being struck with amazement at this, and entreating him to make good this problem by actual experiment, and show some great weight moved by a small engine, he fixed accordingly upon a ship of burden out of the king's arsenal, which could not be drawn out of the dock without great labour and many men; and, loading her with many passengers and a full freight, sitting himself the while far off, with no great endeavour, by only holding the head of the pulley in his hand and drawing the cords by degrees, he drew the ship in a straight line, as smoothly and evenly as if she had been in the sea.

The compound pulley described here was one of Archimedes' inventions. In this passage of Plutarch we also find one version of the famous saying attributed to Archimedes by Pappus: "Give me a place to stand, and I shall move the earth." As we shall see, this invention falls in well with his theoretical studies on mechanics.

Plutarch continues his story of Archimedes' demonstration by telling how Hiero, much impressed, asked him to make war-engines designed both for offense and defense. These were made and found good use under Hiero's successor and grandson Hieronymus in the defense against the Romans under Marcellus. Plutarch has a most dramatic description of the effectiveness of these machines, both for short and long ranges, and for land as well as for sea. At last the Romans became so terrified that "if they but see a little rope or a piece of wood from the wall, instantly crying out, that there it was again, Archimedes was about to let fly some engine at them, they turned their backs and fled." Marcellus laid a long siege to the city, and it was finally taken. Marcellus tried to restrain his soldiers as much as he could from pillaging and looting, and was grieved to see how little he was heeded.

But nothing afflicted Marcellus so much as the death of Archimedes, who was then, as fate would have it, intent upon working out some problem by a diagram, and having fixed his mind alike and his eyes upon the subject of his speculation, he never noticed the incursion of the Romans, nor that the city was taken. In this transport of study and contemplation, a soldier, unexpectedly coming up to him, commanded him to follow to Marcellus; which he declining to do before he had worked out his problem to a demonstration, the soldier, enraged, drew his sword and ran him through. Others write that a Roman soldier, running upon him with a drawn sword, offered to kill him; and that Archimedes, looking back, earnestly besought him to hold his hand a little while, that he might not leave what he was then at work upon inconclusive and imperfect; but the solider, nothing moved by his entreaty, instantly killed him. Others again relate that, as Archimedes was carrying to Marcellus mathematical instruments, dials, spheres, and angles, by which the magnitude of the sun might be measured to the sight, some soldiers seeing him, and thinking that he carried gold in a vessel, slew him. Certain it is that his death was very afflicting to Marcellus; and that Marcellus ever after regarded him that killed him as a murderer; and that he sought for his kindred and honoured them with signal favours.

Plutarch gives here three versions of Archimedes' death, and the farther away from the event we get, the more dramatic the story becomes. In Tzetzes and Zonaras we find the variant that Archimedes, drawing in the sand, said to a Roman soldier who came too close: "Stand away, fellow, from my diagram" which so infuriated the soldier (who, soldier fashion, wouldn't take nothing from nobody) that he killed him. This is the origin of the modern version: "Do not disturb my circles."

This is one of the few episodes of high drama in the history of mathematicians. Much later we find Galois† frantically trying to write down his truly inspired ideas the night before the duel which, as he had feared, proved fatal to him. He was 21 years old. A few mathematical geniuses, for example, the Norwegian Niels Henrik Abel,‡ died of consumption,

† The French mathematician, Évariste Galois (1811–32), showed in the paper written the night before the duel that a general equation of fifth or higher degree cannot be solved by radicals in terms of the coefficients.

‡ The Norwegian mathematician, Niels Henrik Abel (1802–29), found this same result for the fifth degree equation, but in quite a different manner, in 1824.

young and poor. And Condorcet, for one, met with a violent end after the French revolution. But in general mathematicians have been a pretty dull lot in this respect, compared to poets.

Archimedes became, I think, a popular image of the learned man much as Einstein did in our day, and many stories of absent-mindedness were affixed to his name. Thus we read in Plutarch that he would become so transported by his speculations that he would "neglect his person to that degree that when he was occasionally carried by absolute violence to bathe or have his body anointed, he used to trace geometrical figures in the ashes of the fire, and diagrams in the oil on his body, being in a state of entire preoccupation, and, in the truest sense, divine possession with his love and delight in science."

We also have the tale of how he, during one of his (perhaps enforced) baths, discovered the law of buoyancy still known by his name; it excited him so that he ran naked through the streets of Syracuse shouting "Heureka, heureka", which is Greek for "I have found it, I have found it". This story is found, in what I think is a slightly garbled version, in Vitruvius. This discovery enabled Archimedes to confirm Hiero's suspicion that a goldsmith, who had had Hiero's crown or golden wreath to repair, had perpetrated a fraud by substituting silver for gold. Archimedes could now, by weights, determine the crown's density, and he found it smaller than that of pure gold.

These stories of absent-mindedness appeal to our sense of the ridiculous, but it must not be forgotten that a necessary faculty for being a genius of Archimedes' order is a capacity for focusing one's entire attention on the problem at hand for a goodly time to the exclusion of everything else.

This is in essence what we know of Archimedes' life, except for his works. Some traits of his personality, though, can be gleaned from his prefaces and the tales about him; thus we catch a couple of glimpses of a baroque sense of humour. We sense it in his obvious delight in the dramatic demonstration on the beach. And in his preface to his treatise *On Spirals* he tells us that it has been his habit to send some of his theorems to his friends in Alexandria, but without demonstrations, so that they themselves might have the pleasure of discovering the proofs. However, it had annoyed Archimedes that some had adopted his theorems, perhaps as their own, without bothering to prove them, so he tells that he included in the last set of theorems two that were false as a warning "how those who claim to discover everything, but produce no proofs of the same, may be confuted as having actually pretended to discover the impossible".

3.2 Archimedes' Works

While Euclid's *Elements* was a compilation of his predecessors' results, every one of Archimedes' treatises is a fresh contribution to mathematical knowledge.

The works preserved in the Greek are (in probable chronological order):

> On the Equilibrium of Plane Figures, I
> Quadrature of the Parabola
> On the Equilibrium of Plane Figures, II
> On the Sphere and Cylinder, I, II
> On Spirals
> On Conoids and Spheroids
> On Floating Bodies, I, II
> Measurement of a Circle
> The Sand-reckoner

The Greek text of these works was edited in definitive form by J. L. Heiberg. In 1906 he discovered the Greek text of yet another book, *The Method*, hitherto considered lost. It was found in the library of a monastery in Constantinople; the text was written on parchment in a tenth century hand and had been washed off to make the precious parchment available for a book of prayers and ritual in the thirteenth century. Such a text, washed off and with new writing on top of it, is called a *palimpsest* (from a Greek term meaning re-scraping), and is naturally most difficult to read. Luckily, Heiberg could make out enough of this palimpsest to give us a good edition of most of this remarkable book of Archimedes as well as of other treatises of his hitherto poorly preserved or authenticated, among them *The Stomachion*, which has to do with a mathematical puzzle. *The Method* is probably the latest of his preserved works and belongs at the bottom of the above list.

Through Heiberg's sober account of his discovery there shines his joy and pride in this rare find which came as a well-earned reward to a brilliant and diligent scholar.

In addition to these Greek texts there are a couple of treatises that have been preserved, in somewhat corrupted form, through the Arabic translations of Thabit ibn Qurrah (836–901). One is called the *Book of Lemmas* and is available in a Latin version included in Heiberg's edition; it is from this book that the angle-trisection to be presented below is taken. The other was discovered by Schoy and published in 1927; from it we shall take the heptagon construction.

T. L. Heath translated Heiberg's text into English, introducing modern mathematical notation, and this version [9] is now readily available.

In addition to these works that have been preserved we know the titles of several treatises that are lost. Thus we are told of Archimedes' ingenious machine representing the motions of sun, moon, and celestial bodies, and that he even wrote a book on the construction of such devices called *On Sphere-making*.

In order to convey some impression of the nature and scope of Archimedes' achievements I shall describe the contents of his books, though it be only briefly and incompletely.

In the books *On the Equilibrium of Plane Figures* he first proves the law of the lever from simple axioms, and later puts it to use in finding the centres of gravity of several lamina† of different shapes (the notion of centre of gravity is an invention of his). This treatise and his books on floating bodies are the only non-elementary writings from antiquity on physical matters that make immediate sense to a modern reader. Book I of *On Floating Bodies* contains, as Propositions 5 and 6, Archimedes' law of buoyancy, clearly stated and beautifully justified.

But most of Archimedes' books are devoted to pure mathematics. The problems he takes up and solves are almost all of the kind which today call for a treatment involving differential and integral calculus. Thus he finds, in *On the Sphere and Cylinder*, that the volume of a sphere is two-thirds of that of its circumscribed cylinder, while its surface area is equal to the area of four great circles.

Figure 3.1

† One may think of simple plane geometrical figures stamped out of a very thin sheet of metal.

In *The Measurement of the Circle* he first proves that *the area A of a circle of radius r is equal to that of a triangle whose base is equal to the circumference C of the circle and whose height is r, or*

$$A \;\; = \;\; \tfrac{1}{2}rC.$$

From this it follows that the ratio of the area of the circle to the square of its radius is the same as the ratio of its circumference to its diameter. This common ratio is what we call π today, and Archimedes proceeds to calculate that

$$3\tfrac{10}{71} \;\; < \;\; \pi \;\; < \;\; 3\tfrac{10}{70}$$

by computing the lengths of an inscribed and a circumscribed regular polygon of 96 sides. His upper estimate of π is, of course, the commonly used approximation 22/7.

In his book *On Spirals* he studies the curve which we appropriately call *Archimedes' spiral*; see Figure 3.1. If a ray from O rotates uniformly about O, like the hand of a clock, and a point P moves uniformly along this line, starting at O, then P will trace out a spiral of this sort. Its equation in modern polar coordinates† is

$$r \;\; = \;\; a \cdot \theta, \qquad\qquad \theta \geq 0.$$

He finds many surprising properties of this curve, among them the following: Let the curve in Figure 3.2 from O to A be the first turn of an Archimedes' spiral (i.e., corresponding to $0 \leq \theta \leq 2\pi$); the area bounded by this curve and the line segment OA is then one-third of the circle of radius OA. Further, if AB is tangent to the spiral at A, and OB is perpendicular to OA, then OB is equal to the circumference of the circle of radius OA. Though Archimedes does not state it explicitly, this implies that the area of triangle OAB is equal to the area of the circle of radius OA, as we can see using the above theorem from *The Measurement of the Circle*; thus Archimedes has succeeded in both *rectifying* and *squaring* the circle, albeit with fairly complex means. [*Rectifying* a curve (in this case a circle) means to determine a straight line segment of the same length as the curve; *squaring* a figure means determining a square of area equal to that of the figure.]

† The position of a point P in the plane, ordinarily given by Cartesian coordinates x, y, relative to fixed axes, can also be given by so-called *polar coordinates* r, θ, relative to a fixed point O called the pole, and a fixed ray from O, say the positive half of the x-axis. Here r denotes the distance between P and O, and θ denotes the angle from the x-axis to the ray OP measured counterclockwise.

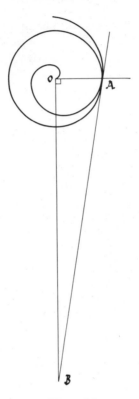

Figure 3.2

In *The Quadrature of the Parabola* he proves the theorem that the area of a segment of a parabola† is four-thirds that of its inscribed triangle of greatest area, a theorem of which he is so fond that he gives three different proofs for it. The *Sand-reckoner*, which he addresses to Gelon, King Hiero's son, is a more popular treatise. In it he displays a number notation of his invention particularly well suited for writing very large numbers. To put this notation to a dramatic test he undertakes to write a number (10^{63}) larger than the number of grains of sand it would take to fill the entire universe, even a universe as large as the one Aristarchos assumed. Aristarchos had proposed a heliocentric planetary system, where the earth travels about a fixed sun once in one year; so in order to explain that the fixed stars apparently keep their mutual distances unchanged during the year, he was forced to maintain

---

† A segment of a parabola is a figure bounded by an arc of a parabola and a straight line. Any line that intersects a parabola and is not parallel to its axis of symmetry cuts off such a segment.

that the fixed star sphere was exceedingly much larger than had commonly been assumed. Here Archimedes furnishes one of our few sources of early Greek astronomy, and he even mentions his own endeavours at measuring the apparent diameter d of the sun. (His estimate is $90°/200 < d < 90°/164$; indeed, the commonly used rough approximation is $d \sim \frac{1}{2}°$.)

The recently discovered *Method* probably belongs at the end of a chronological list of Archimedes' works. In it he applies a certain mechanical method as he calls it—it is closely related to our integration—to a variety of problems with impressive results. This method does not carry the conviction of a proof in his eyes, but is more in the nature of plausibility arguments. He rightly emphasizes the usefulness of such arguments in surmising and formulating theorems which it will be worthwhile to try to prove rigorously. Our last sample of Archimedes' mathematics is taken from *The Method*.

This superficial and incomplete survey of some of Archimedes' works may give some impression of his breadth, originality, and power as a mathematician. A presentation of one of his remarkable chains of proofs in sufficient detail to do it justice lies well beyond the limits I have set myself in this book. I can only hope that a reader whose curiosity about this greatest contribution to ancient mathematics has been aroused will consult the works themselves.

Before we turn to the samples of Archimedes' briefer works it should be mentioned that part of his fame in antiquity rested on his inventions. The compound pulley he demonstrated on the beach and employed in his war-machines is quite a natural consequence of his great theoretical insight into the branch of mechanics now called statics. He furthermore invented the endless screw; this device was (and still is) used, for example, to lift water. Finally his achievements as a mechanical engineer include the construction of the astronomical machine mentioned earlier and, of all things, a hydraulic organ.

3.3 Constructions of Regular Polygons

In modern times the term "construction" in plane geometry has become virtually synonymous with *construction with a pair of compasses and a straightedge*.† In such a construction the rules of the game permit

---

† It is well to preserve the distinction between *ruler* and *straightedge*; a *ruler* is provided with a scale, while a *straightedge* is unmarked.

only the following three operations:

1. A circle may be drawn with any known point as centre and any known line segment as radius.
2. Any two known points may be joined by a line segment.
3. A known line segment may be extended as far as one pleases.

("Known" is here taken to mean either given *a priori* or already constructed by the permitted operations.)

These are severe restrictions, and it is rather surprising that we can achieve as much as we do with such limited means.

Investigations in constructibility have, during the last couple of centuries, taken two directions: (i) There has been a search for equivalent constructions. (By *equivalent constructions* we mean constructions with a new set of tools and permitted operations such that everything that could be constructed with compasses and straightedge and the above rules of the game can also be constructed with the new tools and the new rules; and conversely, everything that can be constructed with the new tools and permitted operations can also be constructed with compasses, straightedge and the old rules.) (ii) Efforts have been made to characterize what can, and particularly what cannot, be constructed with compasses and straightedge by means of the operations enumerated above.

Along the first line it was shown early in the seventeenth century that everything constructible by compasses and straightedge can be constructed with straightedge and a pair of compasses with fixed opening (rusty compasses). Clearly, operation 1 has now been restricted; only a circle with fixed radius may be drawn about any known point. In 1797 the Italian Mascheroni published a book, *Geometria del Compasso*, wherein he showed that with compasses alone one may construct all the points which are constructible with compasses and straightedge. Since two points determine a line, we may consider a line segment constructed if its end points are known; in general, a polygon is considered constructed if its vertices are known, for it is of course impossible to draw the sides with compasses alone. The surprising result that one may dispense with the straightedge without admitting any new operations won fame for Mascheroni, but in 1927 it was discovered that he had been anticipated in 1672 by the Dane Georg Mohr in *Euclides Danicus*, a book which had gathered dust in many libraries for a couple of centuries.

During the nineteenth century several other constructions equivalent to construction by compasses and straightedge were discovered, e.g. construction by straightedge and one fixed circle with known centre, and construction by straightedge and two intersecting fixed circles. If you have lost your compasses, but have a half dollar with which to draw two intersecting circles, everything is all right. But if you do not have a coin you are out of luck, for it is readily shown that straightedge alone does not suffice.†

The other line of investigation—the characterization of possible and impossible constructions—gives rise to considerations of greater mathematical interest. I shall quote but one theorem, due in essence to the great German mathematician C. F. Gauss (1777–1855). It is concerned with the question: which regular polygons are constructible? The theorem says:

A regular polygon of n sides can be constructed by compasses and straightedge if, and only if, either $n = 2^\alpha$ or

$$n = 2^\alpha \cdot p_1 \cdot p_2 \cdots p_r,$$

where p_1, p_2, $\cdots$, p_r are different prime numbers of the form

$$p = 2^{2^\beta} + 1,$$

and α and β are integers ≥ 0.

Some comments on this theorem are in order. Its proof is far too complicated to be given here; suffice it to say that Gauss saw the connection between the geometrical problem of dividing a circle into n equal parts and the algebraic problem of solving the equation

$$x^n = 1;$$

for, the n roots of this equation, when plotted in the complex plane, form the vertices of a regular n-gon inscribed in the unit circle.

But let us see what the theorem says. The factor 2^α is easy to understand, for if we can construct a certain regular polygon, we can immediately construct one with twice as many sides, simply by bisecting all the arcs on the circumscribed circle. From the 15-gon, constructed by Euclid, we can thus immediately get the 30, 60, 120-gon, etc.

The primes of the form

$$2^{2^\beta} + 1$$

were already famous before Gauss discovered their role in this construction problem. They are called *Fermat primes* after the French mathematician Pierre Fermat (1601–1665), the founder of modern number theory, who made the unhappy suggestion that any number of this form

† See, for example, [6], pp. 177 ff.

is a prime. When β takes the values

$$0, \quad 1, \quad 2, \quad 3, \quad 4,$$

$2^{2^\beta} + 1$ takes the values

$$3, \quad 5, \quad 17, \quad 257, \quad 65537,$$

and all these are indeed primes. But in 1735 Euler (1707–1783) found that

$$2^{2^5} + 1 \; = \; 641 \cdot 6{,}700{,}417.$$

We shall show that $2^{2^5} + 1$ is divisible by 641 by writing $2^{2^5} + 1$ as a difference of two integers each of which has 641 as a factor.[†]
On the one hand,

$$641 \; = \; 5^4 + 2^4$$

and hence 641 is a divisor of

$$A \; = \; 2^{28}(5^4 + 2^4) \; = \; 5^4 \cdot 2^{28} + 2^{32}.$$

On the other hand,

$$641 \; = \; 5 \cdot 2^7 + 1$$

and therefore 641 is a divisor of

$$(5 \cdot 2^7 + 1)(5 \cdot 2^7 - 1) \; = \; 5^2 \cdot 2^{14} - 1,$$

and hence also of

$$B \; = \; (5^2 \cdot 2^{14} + 1)(5^2 \cdot 2^{14} - 1) \; = \; 5^4 \cdot 2^{28} - 1.$$

It follows that 641 divides

$$A - B \; = \; 2^{32} + 1.$$

In 1880 Landry showed that

$$2^{2^6} + 1 \; = \; 274{,}177 \cdot 67{,}280{,}421{,}310{,}721.$$

Since then, investigations of these numbers have continued, lately aided by electronic computers. So far, no Fermat primes other than the five listed above have been found. Fermat's conjecture is thus thoroughly exploded, and several mathematicians now tend to believe that these five are all the Fermat primes that exist.

We may check Euclid's results against this theorem. He devised constructions with compasses and straightedge for the regular triangle and

---

[†] See, for example, G. H. Hardy and E. M. Wright: *Introduction to the Theory of Numbers*, 4th ed. Oxford, 1960, Section 2.5, pp. 14–15.

the regular pentagon, and 3 and 5 are the first two Fermat primes. Moreover he constructed the 15-gon, which corresponds to $\alpha = 0$, $r = 2$, $p_1 = 3$, and $p_2 = 5$ in the theorem. No progress was made until Gauss found a construction for the regular 17-gon. It was this construction, of which Gauss was very proud, that set him on the track of the general theorem.

This theorem, it should be noted, implies that, in general, the trisection of an angle by compasses and straightedge is impossible. For, if trisection were possible (say of an angle of 60° or 120°), we could construct a 9-gon from the equilateral triangle; but $9 = 3 \cdot 3$ does not consist of *different* prime factors of the Fermat type, and the 9-gon is hence inconstructible.

It is a common misconception that the Greeks limited themselves entirely to constructions with compasses and straightedge. It is true that virtually all of Euclid's constructions can be performed with these means, but otherwise Greek geometers recognized no such restrictions in their works. Thus we shall see below that Archimedes devised elegant procedures both for the trisection of any angle and for the construction of the regular heptagon (7-gon).

3.4 Archimedes' Trisection of an Angle

We find this construction as Proposition 8 in the *Book of Lemmas* or *Liber Assumptorum* which is preserved in a Latin translation of Thabit ibn Qurrah's Arabic version. The proof below is a slight variation on that in the text, yet the idea is the same.

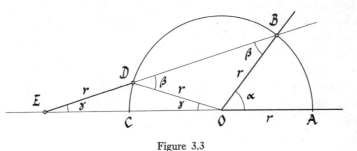

Figure 3.3

In Figure 3.3, let α be any given angle; its vertex is called O. On one of its sides we choose a point A, and with O as a centre and $OA = r$ as a radius we draw a circle; let it intersect α's other side at B, and let A's diametrically opposite point be C. OC is produced beyond C.

Now comes the crucial construction. On our straightedge we make two marks so that the distance between them is r; call the left-hand mark L and the right-hand mark R. (Note that we have just violated the usual rules for compasses and straightedge construction because we have marked the straightedge.) We now place the straightedge so that it passes through B and mark R lies on the arc CB of the circle. Then we move the straightedge in such a way that mark R travels along the circle and the straightedge always passes through B until the mark L falls on the extension of OC. Line BDE represents this position of the straightedge, that is, it passes through B and $DE = r$.

We shall now show that the angle γ at E is one-third of α.

We first observe that the angle β at D is twice γ, for it is supplementary to $\angle D$ in the isosceles triangle ODE, and is thus equal to the sum 2γ of the remaining two angles. Now α is supplementary to $\angle O$ in triangle EOB and is hence equal to the sum $\beta + \gamma$ of the two remaining angles. We thus have:

$$\alpha \;=\; \beta + \gamma \;=\; 2\gamma + \gamma \;=\; 3\gamma.$$

So γ is a third of α, and we have succeeded in trisecting the given angle. Since this cannot be done in ordinary constructions with compasses and straightedge, we must have used an unpermitted operation; it consisted, as noted earlier, in placing two marks on our straightedge. This enabled us to fit a line segment between two given curves (here the semicircle and a straight line) while its extension passes through a given point (here B). This operation is a powerful addition to the three permissible operations in ordinary constructions, and it makes it possible to solve several new problems, for example the one we saw here. This operation was not new in Greek mathematics at the time of Archimedes; a construction utilizing it even had a special name: it was called a *neusis-construction*, from the Greek verb *neuein*, meaning to nod or verge (the marked line segment on our ruler "verges" or points towards the fixed point B).

Although Archimedes' authorship of this proof has been questioned (we do not have the Greek text), I have no doubt that the theorem is his, for it resembles intimately some of his theorems in *On Spirals*.

Problems

3.1 Given an angle α $(= \angle AOB$; see Fig. 3.4), on one side of α we mark off an arbitrary line segment $OA = a$. Through A we draw two lines, one parallel and one perpendicular to the other side of α. A line segment CD of length $2a$ is fitted between these two lines so that its extension beyond C passes through O (a *neusis* construction). Show that, in terms of the notations on Figure 3.4,

$$\alpha \; = \; 3\beta,$$

i.e., that α has been trisected. This elegant construction is found in the works of Pappus (ca. 300 A.D.), and it may well be due to Pappus himself.

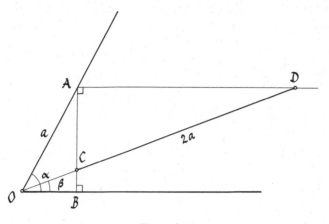

Figure 3.4

3.2 A tomahawk-shaped plate (Figure 3.5) is jammed into an arbitrary angle α so that the upper edge of the ruler passes through its vertex. Show that

$$\alpha \; = \; 3\beta$$

or that the tomahawk is a simple angle-trisector. I do not know the origin of this lovely device. It was communicated to me by a student.

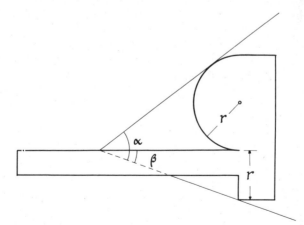

Figure 3.5

3.5   Archimedes' Construction of the Regular Heptagon

In the course of his investigations on Islamic mathematics, Carl Schoy found a manuscript of an Arabic translation by Thabit ibn Qurrah of an otherwise unknown treatise by Archimedes. It was published in 1927 after Schoy's death. This treatise contains as its sixteenth and last proposition Archimedes' construction of the regular heptagon which I shall present below, following Tropfke's German edition.[†]

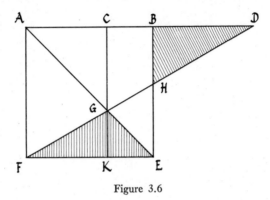

Figure 3.6

(a) We begin the construction in Figure 3.6 with a given line segment AB. On AB we construct a square $AFEB$, draw its diagonal AE, and extend the given line segment AB beyond B. We now perform a *neusis* construction: we let a line vary so that it always passes through F until the area between this line, the diagonal AE, and the side FE is equal to the area between the line, the side BE, and the extension of AB beyond B. FD represents this position of the line, and the equal areas are shaded in the figure. FD intersects the diagonal at G, the side BE at H, and the extension of AB at D. Through G we draw a line parallel to BE intersecting AB at C and FE at K. We now maintain that the four collinear points A, B, C, D satisfy the following two equations:

(i) $$AB \cdot AC = BD^2$$

(ii) $$CD \cdot CB = AC^2.$$

This is proved as follows. From the equality of the two shaded triangles we obtain

$$GK \cdot FE = BH \cdot BD$$

or

† J. Tropfke, "Die Siebeneckabhandlung des Archimedes," *Osiris* 1, pp. 636–651.

$$(1) \qquad \frac{BH}{GK} = \frac{FE}{BD}.$$

Triangle HBD is similar to triangle GKF, for they are both right, and the angles at F and D are equal. Hence we have

$$(2) \qquad \frac{BH}{GK} = \frac{BD}{FK}.$$

Equations (1) and (2) yield

$$\left(\frac{BH}{GK} =\right) \frac{FE}{BD} = \frac{BD}{FK},$$

from which we get

$$FE \cdot FK = BD^2.$$

Substituting AB for FE and AC for FK, we obtain (i) above, i.e.

$$AB \cdot AC = BD^2.$$

We now note that triangle FKG is similar to triangle DCG and get

$$\frac{GK}{FK} = \frac{GC}{CD}$$

or

$$(3) \qquad GK \cdot CD = FK \cdot GC.$$

Since AE is a diagonal in a square, $\sphericalangle GAC$ and $\sphericalangle GEK$ are each $45°$; hence $GC = AC$ and $GK = KE$. Further, $FK = AC$ and $KE(=GK) = CB$. Substituting CB for GK and AC for FK and GC in (3) we get

$$CB \cdot CD = AC^2$$

which is, indeed, (ii) above.

At this point the impatient reader may well wonder what this has to do with a heptagon. We shall bring out the connection after recalling a theorem which we shall use frequently below (see Figure 3.7): *The locus of all points P on one side of a given line segment AB, such that angle APB is a given angle α, is a circular arc determined by the chord AB and the angle α; $\sphericalangle APB = \alpha$ is measured by half the circular arc on the other side of AB.*†

† For details on how to construct this locus, see, e.g., *Hungarian Problem Book I*, p. 30, NML 11.

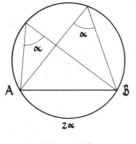

Figure 3.7

(b) In Figure 3.8, AD is the line divided as in Figure 3.6, that is, so that

(i) $$AB \cdot AC \ = \ BD^2$$

(ii) $$CD \cdot CB \ = \ AC^2.$$

We now construct E so that

$$CE \ = \ CA$$

and

$$BE \ = \ BD.$$

We further construct the circumscribed circle of triangle AED (this is the large circle in our figure). We now assert that AE is the side of the regular heptagon inscribed in this circle.

The proof of this claim is not simple; yet it is less complicated than the result is astonishing.

We begin by observing that since triangle EBD is isosceles, the angles α at its base are equal. Similarly for the angles at the base of triangle ACE; they are marked β.

We now extend EB and EC until they meet the circle at F and G, respectively. We draw AF and let H be its intersection with EG. We join H and B.

We recall that angles with their vertices on the circumference of a circle are half of the arc they subtend. Therefore each of the minor arcs AE and DF is 2α; hence $\sphericalangle FAD = \alpha$ and $\sphericalangle AFE = \alpha$. They are so marked.

We note that $\sphericalangle EBA$ is supplementary to $\sphericalangle B$ in triangle EBD. It is therefore equal to the sum 2α of the other two angles. It is so marked.

We shall now use our condition (ii). It implies

$$\frac{CD}{AC} \ = \ \frac{AC}{CB}$$

or, since $AC = EC$,

$$\frac{CD}{EC} = \frac{EC}{CB}.$$

This means that triangles BEC and EDC are similar, for they have the angle at C in common and a pair of corresponding sides proportional. Hence $\sphericalangle BEC = \alpha$ (it is so marked) and the minor arc GF is then 2α and is equal to the arcs AE and DF.

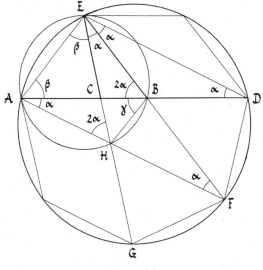

Figure 3.8

The arcs ED and AG are equal, for each is 2β. If we can show that $\beta = 2\alpha$ we shall have finished the proof, for then we shall have that the arc AE is one-seventh of the entire circumference. In order to do this we observe that since the segment HB subtends the angle α at A and at E, A and E must both lie on a circular arc with HB as a chord. In other words, the quadrilateral $AHBE$ is inscribable, and its circumscribed circle is drawn. In this circle the peripheral angles β subtend the chords EB and AH and so these chords are equal. Further, the angle at H subtending the chord AE is equal to the angle marked 2α at B subtending the same chord; $\sphericalangle AHE$ is consequently also marked 2α.

We now make use of our condition (i) in the form

$$\frac{AB}{BD} = \frac{BD}{AC}$$

or, since $EB = BD = AH$,

$$\frac{AB}{AH} = \frac{EB}{EC}.$$

This implies that triangles EBC and ABH are similar since, in addition,

$$\sphericalangle BAH = \sphericalangle BEC = \alpha.$$

Thus the angle marked γ on the figure is equal to 2α. This concludes our long proof, for γ and β are both opposite the chord AH in the small circle. Hence, $\beta = 2\alpha$, and arc ED, as well as arc AG, is 4α. The total circumference of the large circle is then 14α, and arc AE is a seventh of the circumference. The regular heptagon is drawn in the figure.

The neusis construction in part (a) above is unique in Greek mathematics, and one may feel that it is intuitively less satisfactory than the sort used in the trisection. Indeed, I do not know precisely how Archimedes proposed to decide when equality of the two triangles is achieved, and the Arabic text certainly contains no clue. However, it can readily be shown that the equations (i) and (ii) can be solved by intersecting two conic sections, a method widely used by Greek geometers.

Problem

3.3 Show that, if we set $AB = a$, $AC = x$, $BD = y$, then (i) and (ii) reduce to

$$\text{(i) } y^2 = ax, \quad \text{(ii) } y = \frac{a^2}{a-x} - 2a.$$

If (x, y) denote Cartesian coordinates, (i) represents a parabola and (ii) an hyperbola. Sketch the two curves carefully and show graphically that there is but one acceptable solution of (i) and (ii).

The Greek mathematicians were perfectly capable of reducing the solution of a pair of second degree equations in two unknowns to the problem of intersecting two conics, even though they did not have our convenient algebraic notation at their disposal.

3.6 Volume and Surface of a Sphere According to *The Method*

If we rotate Figure 3.9 about the dotted line, we generate a cone inscribed in a hemisphere which, in turn, is inscribed in a cylinder. The volumes of these three figures have the ratio $1:2:3$. This beautiful theorem is a variant of Archimedes' favorite result. He was, in fact, so proud of it that he wanted a sphere with its circumscribed cylinder and their ratio (2:3) engraved on his tombstone. He got his wish, as we know from Cicero, who, when quaestor in Sicily, found Archimedes' tomb in a neglected state and restored it.

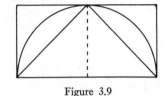

Figure 3.9

To prove this theorem in a rigorous and unexceptionable fashion is one of the chief aims of the first book of *On the Sphere and Cylinder* (the other is to demonstrate that the surface of the sphere is four great circles). The reader of this book is forcibly impressed by the elegance of the sequence of theorems leading him through surprising and dramatic turns to the two final goals, but at the same time he cannot help recognizing that this sequence surely does not map the road which Archimedes first followed to discover these results.

When he writes this way, Archimedes is merely following the common practice of the Greek geometers—indeed of most polished mathematical writing—which aims at convincing the reader of the validity of certain results and not at teaching him how to discover new theorems on his own.

This lack of the analytic and heuristic element in codified Greek geometry, i.e., of open display of the way in which theorems were first surmised rather than proved, was deplored in the seventeenth century when mathematicians were striving to create a new mathematical analysis (calculus and its ramifications). The English mathematician Wallis (1616–1703) even went so far as to believe that the Greeks deliberately had hidden their avenues of discovery.

Here we have one of a great many instances where lack of textual material has led modern scholars to false conclusions, for Wallis' surmise was thoroughly disproved when Heiberg found Archimedes' *Method*. Its aim is well described in the preface dedicating it to Eratosthenes. Archimedes writes here, in part (in Heath's translation):

> ... I thought fit to write out for you and explain in detail in the same book the peculiarity of a certain method, by which it will be possible for you to get a start to enable you to investigate some of the problems in mathematics by means of mechanics. This procedure is, I am persuaded, no less useful even for the proof of the theorems themselves; for certain things first became clear to me by a mechanical method, although they had to be demonstrated by geometry afterwards because

their investigation by the said method did not furnish an actual demonstration. But it is of course easier, when we have previously acquired, by the method, some knowledge of the questions, to supply the proof than it is to find it without any previous knowledge. This is a reason why, in the case of the theorems the proof of which Eudoxus was the first to discover, namely that the cone is a third part of the cylinder, and the pyramid of the prism, having the same base and equal height, we should give no small share of the credit to Democritus who was the first to make the assertion with regard to the said figure though he did not prove it. I am myself in the position of having first made the discovery of the theorem now to be published by the method indicated, and I deem it necessary to expound the method partly because I have already spoken of it and I do not want to be thought to have uttered vain words, but equally because I am persuaded that it will be of no little service to mathematics; for I apprehend that some, either of my contemporaries or of my successors, will, by means of the method when once established, be able to discover other theorems in addition, which have not yet occurred to me.

In the following I shall present Archimedes' argument from mechanics which led him to the theorem about the cone, the sphere, and the cylinder (*The Method*, Proposition 2).

I should say first that part of the theorem was known before Archimedes; he tells himself, in the above excerpt, that Eudoxos proved that a cone is one-third of its circumscribed cylinder, a theorem which we know as Euclid **XII**, 10. (We have here an instance, and one of the most important ones, where on ancient authority we can assign part of the *Elements* to one of Euclid's predecessors.) Archimedes' job is therefore to relate the sphere either to the cylinder or to the cone.

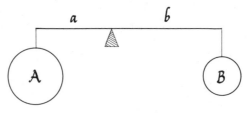

Figure 3.10

We shall need the law of the lever: The lever (see Figure 3.10) is in equilibrium if the product of the weight A and the distance a between the fulcrum and the point of suspension of A is equal to the product of the weight B and its distance b from the fulcrum. In symbols

$$A \cdot a \;=\; B \cdot b.$$

Archimedes actually prefers to write the condition for equilibrium in the form

$$\frac{A}{B} = \frac{b}{a},$$

and so avoids several embarrassing points such as a physical interpretation for the quantity $A \cdot a$ (called *moment*). This equilibrium condition is found as Propositions 6 and 7 in *On the Equilibrium of Plane Figures*, I.

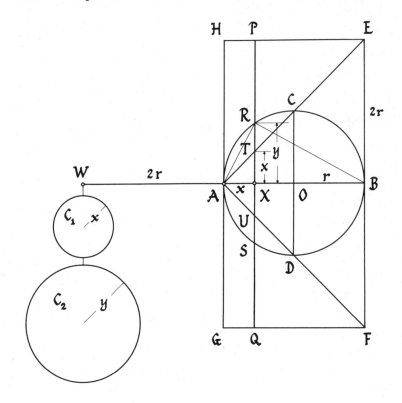

Figure 3.11

In the circle with centre O and radius r (see Figure 3.11) two perpendicular diameters AB and CD are drawn. AC is drawn and extended to E, where it intersects the perpendicular at B to AB; and AD is drawn and extended to its intersection with EB at F. We note that

$$EB = 2r = BF,$$

since angles BAE and BAF are each 45°. The rectangle $EFGH$ on EF with altitude AB is completed.

PQ is a line perpendicular to AB through an arbitrary point X on AB. It intersects the circle at R and S, AE at T, and AF at U. We set

$$XT \ = \ x$$

and

$$XR \ = \ y$$

and observe that also

$$AX \ = \ x.$$

From the right triangle AXR we see that

(1) $x^2 + y^2 \ = \ AR^2.$

From the right triangle ARB (the angle at R is right for it subtends a diameter), we see that

(2) $AR^2 \ = \ x \cdot 2r,$

for a side in a right triangle is the mean proportional between its projection on the hypotenuse and the entire hypotenuse (see Chapter 2, p. 61). Combining (1) and (2) we get:

(3) $x^2 + y^2 \ = \ x \cdot 2r.$

AB is extended beyond A to W so that

$$WA \ = \ 2r,$$

and this ends the construction.

We now rotate the entire figure about WB. The circle thus generates a sphere of radius r; triangle EAF generates a cone whose base is a circle of radius $BE = 2r$, and whose height is $AB = 2r$; and rectangle $EFGH$ generates a cylinder whose base has radius $BE = 2r$ and whose height is $AB = 2r$. During the rotation, the arbitrary line PQ swept out a plane which intersects the cylinder in a circle of radius $2r$, the cone in a circle C_1 of radius x, and the sphere in a circle C_2 of radius y.

Dividing both sides of (3) by $(2r)^2$ we obtain

(4) $\dfrac{x^2 + y^2}{(2r)^2} \ = \ \dfrac{x}{2r}$

or

(5) $\dfrac{\pi x^2 + \pi y^2}{\pi (2r)^2} \ = \ \dfrac{x}{2r}.$

(Introducing π is taking quite a liberty with the text of Archimedes, who argues entirely on the basis of the fact that the ratio of the areas of two circles equals the ratio of the squares of their diameters.)

WB is now considered as a lever supported at A, and equation (5) can then be interpreted as the equilibrium condition in the following way: The ratio of the sum of the areas of circles C_1 and C_2 (cut out of the cone and sphere, respectively) to the area of the circle of radius $2r$ (cut out of the cylinder) is equal to the ratio of the distance x from A to X to the distance $2r$ from A to W. In other words, if we imagine circular discs whose weights are proportional to their areas, then the circles C_1 and C_2 suspended from W would balance the circle of area $\pi(2r)^2$ suspended from X on a lever with fulcrum at A.

This equilibrium condition is satisfied for any position of PQ between HG and EF. If we think of the cylinder, the sphere and the cone as made up of the circles cut in these solids by every possible plane swept out by PQ, and if for each PQ we suspend C_1 and C_2 from W, then all the circles C_1 and C_2, reassembled to form the cone and the sphere, will balance the cylinder. Since the cylinder's centre of gravity is at O, we may express this as

$$\frac{\text{sphere} + \text{cone}}{\text{cylinder}} = \frac{AO}{AW} = \frac{1}{2}$$

or

$$2(\text{sphere} + \text{cone}) = \text{cylinder}.$$

Since the cylinder is thrice the cone, we get

$$2 \cdot \text{sphere} = \text{cone } EAF.$$

But

$$\text{cone } CAD = \tfrac{1}{8} \text{ cone } EAF$$

or

$$\text{cone } EAF = 8 \cdot \text{cone } CAD,$$

for, when the linear scale of a figure is multiplied by $\frac{1}{2}$, its volume is multiplied by $(\frac{1}{2})^3$. Thus

$$\text{sphere} = 4 \cdot \text{cone } CAD,$$

or *a sphere is equal to four times a cone whose base is a great circle of the sphere and whose height is the sphere's radius.*

It is now a simple matter to show that a sphere is two-thirds of its circumscribed cylinder.

Problem

3.4 Complete the proof that the volume of a sphere is two-thirds the volume of its circumscribed cylinder.

Archimedes gets more out of this lovely argument. He says, again in Heath's translation:

> From this theorem, to the effect that a sphere is four times as great as the cone with a great circle of the sphere as base and with height equal to the radius of the sphere, I conceived the notion that the surface of any sphere is four times as great as a great circle in it; for, judging from the fact that any circle is equal to a triangle with base equal to the circumference and height equal to the radius of the circle, I apprehended that, in like manner, any sphere is equal to a cone with base equal to the surface of the sphere and height equal to the radius.

This is the first and one of the finest examples of bold analogy in the history of mathematics, and it will pay to scrutinize this argument.

We begin by noting that the area of a triangle with base b and altitude h can be expressed as the sum of areas of triangles $\triangle_1, \triangle_2, \cdots, \triangle_k$ with common altitude h and with bases $b_1, b_2, \cdots, b_k$ such that $b_1 + b_2 + \cdots + b_k = b$. Similarly, the volume of a pyramid (or a cone) can be expressed as the sum of volumes of pyramids (or cones) with the same height, the areas of whose bases add up to the area of the base of the given pyramid (or cone).

I believe that Archimedes thought of a circle as made up of many "triangles" with their vertices at its centre and their bases along its circumference. Their altitudes are all equal to the radius, so one may add their areas by adding the bases. Thus the area of a triangle with altitude equal to the radius and base equal to the circumference is equal to the area of the circle. This is, indeed, the theorem which Archimedes quotes above and proves rigorously in *The Measurement of the Circle*.

I believe further that this made Archimedes think of a sphere of radius r as made up of many "cones" or "pyramids" with their vertices at its centre and their bases on its surface. All these "cones" have altitude r, so one may add their volumes by adding their bases. Thus it is reasonable to assume that the volume of a sphere equals that of a cone of height r with base equal in area to the sphere's surface.

But the lever argument showed that the volume of a sphere equals four cones of height r and base equal in area to a great circle; so the sphere is equal to the sum of these cones, i.e., to one cone of height r with base equal to four great circles. Thus the sphere is equal to either of two cones of height r; the area of the bases of these cones must then be equal to four great circles. It follows that the surface of the sphere is equal to the area of four great circles.

What Archimedes said at the end of Proposition 1 of *The Method* applies equally well here. He said:

> Now the fact here stated is not actually demonstrated by the argument used; but that argument has given a sort of indication that the conclusion is true. Seeing then that the theorem is not demonstrated, but at the same time suspecting that the conclusion is true, we shall have recourse to the geometrical demonstration which I myself discovered and have already published.

What makes the mechanical procedure unacceptable as a proof in Archimedes' fastidious eyes is the part where solids are considered as sums of plane sections. We are now familiar with such procedures under the name of *integration*; Archimedes however succeeds in shifting the burden of integration over to a determination of the centre of gravity of a cylinder, which is so simple that a mere symmetry consideration suffices.

Archimedes' remark in his preface to *The Method*, that it will be of no little service to mathematics, was truly prophetic. But since this book was lost, the mathematicians of the seventeenth century had to develop the theory of integration on their own; it is futile but tempting to speculate on what influence *The Method* might have had on the course of events. As it happened, it was not until the latter part of the nineteenth century that the theory of integration reached a level of rigour which would have found favour with Archimedes.

These small samples of Archimedes' work are far from doing him justice. They may, however, serve to give some impression of his fearlessness, power, and ingenuity when faced with a difficult problem. In order to gain a proper appreciation of his mastery of his art one must become intimate with at least one of his elegant structures of theorems crowned by a difficult and beautiful result.

CHAPTER FOUR

Ptolemy's Construction of a Trigonometric Table

4.1 Ptolemy and The *Almagest*

Klaudios Ptolemaios, or Ptolemy, lived and worked in Alexandria around 150 A.D. Although the precise dates and details of his life are unknown to us, his principal work, now commonly called the *Almagest*, supplies the evidence for placing him in the middle of the second century, for in this work he quotes his own observations of identifiable astronomical events.†

Ptolemy did some work in pure mathematics, but he is famous as an applied mathematician. (It is, however, doubtful that he would have taken the modern distinction between pure and applied mathematics seriously.) His *Almagest* played the same role in mathematical astronomy as Euclid's *Elements* and Apollonius' *Conics* did in their respective subjects; it made its predecessors utterly superfluous, and so they are practically all lost. But Ptolemy, unlike Euclid, acknowledged his precursors' achievements generously and precisely, so our knowledge of pre-Ptolemaic

---

† Thus we read in *Almagest*, Book IV, Ch. 6: "Of the three lunar eclipses, selected from those which we ourselves have most carefully observed at Alexandria, the first took place in the 17th year of Hadrian (the Roman emperor) on the 20/21 day of the Egyptian month Payni; the middle occurred one half and one quarter equinoctial hour (¾ hour) before midnight; the eclipse was total, and for this hour the true position of the sun was $13\frac{1}{4}°$ of Taurus, very nearly," and similarly for the other two eclipses. This eclipse is readily identified as the one that took place A.D. 133, May 6, 22^h 7^m G.M.T. [Lunar Eclipse No. 2071 in v. Oppolzer: *Canon of Eclipses* (reissued by Dover Publications in 1962)].

This is but one of a host of examples that might be quoted.

101

astronomy is richer and firmer than our knowledge of pre-Euclidean mathematics. For the same reason, we can identify quite well Ptolemy's own contributions.

Mathematical astronomy is, by a wide margin, the oldest exact science. The Babylonians of the Seleucid Era, i.e. of the last three centuries B.C., had already devised elegant schemes yielding good quantitative predictions of astronomical phenomena. It is not surprising to anyone acquainted with Babylonian mathematics that these schemes are entirely arithmetical in nature, without a trace of underlying geometrical models; nor is it surprising that all serious attacks on astronomical problems in the Greek world are based on geometrical models. Indeed, the position of mathematical astronomy among learned subjects was such that, with few exceptions, every great geometer from Eudoxos on devoted some of his efforts to astronomical matters. We are able to trace, with some certainty, the development of geometrical models for the motions of sun, moon, and planets from the qualitative devices of Eudoxos to their culmination in Ptolemy's simple, elegant, and yet quantitatively excellent models set forth in the *Almagest*. (An example of these is given in the appendix.)

The curious name of Ptolemy's greatest work probably comes from an Arabic distortion of a Greek name meaning "the greatest" (hē megistē); Ptolemy's own title is, in translation, *The Mathematical Collection*. In this book he develops not only his astronomical models, but also the mathematical tools beyond elementary geometry needed in astronomy, among them trigonometry. The *Almagest* is a masterpiece of exposition; Ptolemy never presents a table without first explaining how it may be computed, and the parameters of his models are all derived in plain sight from carefully quoted observations.

Several of Ptolemy's minor works have reached us. Whatever their subject—geography, astrology, or harmonics (theory of music)—we recognize in them Ptolemy's genius for methodological arrangement and presentation of his material.

The *Almagest* is a large and technical book; furthermore, the only reliable editions of it are Heiberg's definitive Greek text and its translation into German by Manitius, so the *Almagest* is not widely read nowadays. But ignorance does not keep people from writing books. It is therefore not surprising that the common version of the relation between Ptolemy's and Copernicus' work is a gross distortion, not to Ptolemy's advantage. Actually, the influence of the *Almagest* can hardly be overestimated. More than any other book, it contributed to the idea, so basic in all scientific endeavour, that a quantitative, mathematical description of natural phenomena, capable of yielding reliable predictions, is both possible and desirable.

4.2 Ptolemy's Table of Chords and Its Uses

The theory that provides numerical solutions to geometrical problems involving angles is called *trigonometry*; this means literally *measurement of triangles*. Ptolemy develops this subject in chapters ten and eleven of the first book of the *Almagest*. The eleventh chapter consists of a table of chords, whose beginning and end I have copied and translated in the table shown in Figure 4.1, and the tenth chapter tells how such a table may be computed. We shall look closely at chapter ten in the following section, but first we turn our attention to the table.

Κανόνιον τῶν ἐν κύκλῳ εὐθειῶν			Table of Chords		
περιφε. ρειῶν	εὐθειῶν	ἑξηκοστῶν	arcs	chords	sixtieths
∠ʹ	σ λα κε	σσ α β ν	½°	0;31,25	0;1,2,50
α	α β ν	σσ α β ν	1°	1;2,50	0;1,2,50
α∠ʹ	α λδ ιε	σσ α β ν	1½°	1;34,15	0;1,2,50
β	β ε μ	σσ α β ν	2°	2;5,40	0;1,2,50
β∠ʹ	β λζ δ	σσ α β μη	2¼°	2;37,4	0;1,2,48
γ	γ η κη	σσ α β μη	3°	3;8,28	0;1,2,48
γ∠ʹ	γ λθ νβ	σσ α β μη	3½°	3;39,52	0;1,2,48
δ	δ ια ις	σσ α β μζ	4°	4;11,16	0;1,2,47
δ∠ʹ	δ μβ μ	σσ α β μζ	4½°	4;42,40	0;1,2,47
ε	ε ιδ δ	σσ α β μς	5°	5;14,4	0;1,2,46
ε∠ʹ	ε με κζ	σσ α β με	5¼°	5;45,27	0;1,2,45
ς	ς ις μθ	σσ α β μδ	6°	6;16,49	0;1,2,44
ς∠ʹ	ς μη ια	σσ α β μγ	6½°	6;48,11	0;1,2,43
ζ	ζ ιθ λγ	σσ α β μβ	7°	7;19,33	0;1,2,42
ζ∠ʹ	ζ ν νδ	σσ α β μα	7½°	7;50,54	0;1,2,41
⋮	⋮	⋮	⋮	⋮	⋮
ροδ∠ʹ	ριθ να μγ	σσ σσ β νγ	174½°	119;51,43	0;0,2,53
ροε	ριθ νγ ι	σσ σσ β λς	175°	119;53,10	0;0,2,36
ροε∠ʹ	ριθ νδ κζ	σσ σσ β κ	175½°	119;54,27	0;0,2,20
ρος	ριθ νε λη	σσ σσ β γ	176°	119;55,38	0;0,2,3
ρος∠ʹ	ριθ νς λθ	σσ σσ α μζ	176½°	119;56,39	0;0,1,47
ροζ	ριθ νζ λβ	σσ σσ α λ	177°	119;57,32	0;0,1,30
ροζ∠ʹ	ριθ νη ιη	σσ σσ α ιδ	177½°	119;58,18	0;0,1,14
ροη	ριθ νη νε	σσ σσ σσ νζ	178°	119;58,55	0;0,0,57
ροη∠ʹ	ριθ νθ κδ	σσ σσ σσ μα	178½°	119;59,24	0;0,0,41
ροθ	ριθ νθ μδ	σσ σσ σσ κε	179°	119;59,44	0;0,0,25
ροθ∠ʹ	ριθ νθ νς	σσ σσ σσ θ	179½°	119;59,56	0;0,0,9
ρπ	ρκ σσ σσ	σσ σσ σσ	180°	120;0,0	0;0,0,0

Figure 4.1

The numbers are written with the Greek letters employed as numerals. The Greek letters in their usual order were divided into three groups of nine letters each, corresponding to units, tens, and hundreds (see Figure 4.2). The classical Greek alphabet has only 24 letters, but three obsolete characters are retained and used, in their ancient places, as 6, 90, and 900. The letter representing 6 is, in its capital form, the old *waw*, or *digamma*, as it is usually called because it resembles two gammas. It became, in the lower case form, a sign called *stigma* and which otherwise served as a ligature or contraction of the two letters σ and τ. The letter for 90 is called *qoppa*, and that for 900 *sampi*. Thousands are written as ones but with a stroke in the lower left corner. There were various ways of writing ten thousands, or myriads, but they do not concern us here. The sign for zero is given in two variants at the bottom of Figure 4.2. We know it from Greek papyri written in Ptolemy's time. The common statement that the Greek zero looked like *o*, and was an abbreviation of *ouden*, which is Greek for "nothing", is false. This zero is not found until Byzantine times. In Ptolemy's day *o* could only mean 70.

A	α	1	Ι	ι	10	Ρ	ρ	100	,α 1000
B	β	2	Κ	κ	20	Σ	σ	200	,β 2000
Γ	γ	3	Λ	λ	30	Τ	τ	300	,γ 3000
Δ	δ	4	Μ	μ	40	Υ	υ	400	,δ 4000
E	ε	5	Ν	ν	50	Φ	φ	500	,ε 5000
F	ϛ	6	Ξ	ξ	60	Χ	χ	600	,ϛ 6000
Z	ζ	7	Ο	ο	70	Ψ	ψ	700	,ζ 7000
H	η	8	Π	π	80	Ω	ω	800	,η 8000
Θ	θ	9	Q	ϙ	90	ϡ	ϡ	900	,θ 9000

$$\overset{\circ}{\sigma}, \sigma\, \text{etc}: 0$$

Figure 4.2

In the first column of the table we find still one more sign; it looks like our symbol for angle with an apostrophe and it means $\frac{1}{2}$. This is all we need to know in order to transcribe the table.

The trigonometric tables used today are lists of values of the two basic trigonometric functions, sine and cosine, of an angle α; we write them $\sin\alpha$ and $\cos\alpha$. They are defined (see Figure 4.3) as the opposite and adjacent side, respectively, of the angle α when it is an angle in a right triangle whose hypotenuse is 1. This definition, of course, is valid only for angles less than 90°, but that suffices for our purposes. The two functions are not independent, and we see by Pythagoras' theorem, that

$$\sin^2\alpha + \cos^2\alpha = 1.$$

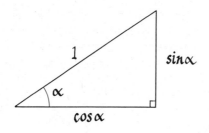

Figure 4.3

Ptolemy tabulates neither of these functions, but the related chord function of the arc α or, as we shall write, crd α. It is defined as *the length of the chord corresponding to an arc of α degrees in a circle whose radius is 60* (see Figure 4.4a). There is a simple relation between the chord and the sine of an angle, for, as we see from Figure 4.4,

$$\text{crd } \alpha \quad = \quad 120 \cdot \sin \frac{\alpha}{2}.$$

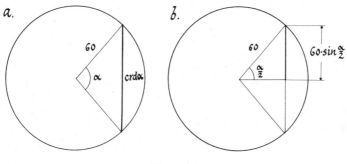

Figure 4.4

In the table of Figure 4.1 we find values of crd α for α going, in steps of $\frac{1}{2}°$, from $\frac{1}{2}°$ to 180°. Ptolemy says that he expresses the chords in sexagesimals "because of the inconvenience of fractions". Thus when the table gives

$$\text{crd } 4\tfrac{1}{2}° \quad = \quad 4;42,40,$$

it means that

$$\text{crd } 4\tfrac{1}{2}° \quad = \quad 4 + \frac{42}{60} + \frac{40}{60^2}$$

measured in units which are 1/60 of the radius.

We recognize here the Babylonian sexagesimal number system, but transcribed into Greek numerals. It is no wonder that Ptolemy adopts it as soon as he has to do serious computations, for it was the only sensible system in antiquity, at least as far as fractions are concerned. Ptolemy is, however, not consistent in his borrowing, for he continues to write the integral parts in the usual Greek fashion. Thus we find in the table:

$$\text{crd } 177\tfrac{1}{2}° \ = \ 119;58,18,$$

where 119 is written $\rho\iota\theta$, while a consistent Babylonian would have written it

$$\text{crd } 2,57;30° \ = \ 1,59;58,18.$$

We are even less consistent when we write 120°30′29.2″ where Ptolemy at least would have written 120°30′29″12‴, to use the standard notation for degrees, minutes, seconds, and thirds.

Such a mixture of elements from different civilizations is characteristic of Hellenistic culture.† We find many other acknowledged borrowings from Babylonia in the *Almagest*, and we can identify several Egyptian elements as well. Thus the division of the day into 24 hours used throughout the *Almagest* is of Egyptian origin, and occasionally we find Ptolemy writing fractions in the Egyptian fashion, such as

$$\angle'\delta' \ = \ \tfrac{1}{2}+\tfrac{1}{4}$$

for $\tfrac{3}{4}$, instead of 0;45.‡

The third column of the table, the one marked "sixtieths", gives 1/30 of the increment from one line to the next. Since the difference in arc from one line to the next is half a degree, or 30 minutes of arc, the third column gives the average increase in crd α corresponding to an increase of one minute in α; one minute is a sixtieth of a degree, whence the heading. This column is used for interpolation, i.e. for finding crd α if α lies between two entries in the arc column. Thus we may find crd 6°32′, or crd 6;32° (correct to the nearest 3600th) in the following way:

$$
\begin{aligned}
\text{crd } 6\tfrac{1}{2}° \ &= \ 6;48,11 \\
2\cdot(0;1,2,43) \ &= \ 0;\ 2,\ 5,26 \\
\hline
\text{crd } 6;32° \ &= \ 6;50,16
\end{aligned}
$$

---

† The term *Hellenistic*, applied to the Greek-speaking world centered on the Eastern Mediterranean, usually refers to the period from Alexander's death in 323 B.C. to the Roman Conquest; but in the city of Alexandria, for one, Hellenistic culture survived far into Roman times.

‡ An example of this practice is found in the quotation in the note on p. 101, where $\tfrac{1}{2}$ and $\tfrac{1}{4}$, however, are spelled out as words.

for, crd 6;32° exceeds crd 6;30° by twice the increment corresponding to one minute. Thus, when an arc is given to the nearest minute we are able to find its chord, to two sexagesimals, in a circle of radius 60. Using the inverse procedure, the table enables us to find an arc whose chord (in a circle of radius 60) is given.

It appears that these two basic operations—to find crd α when α is given, and to find α when crd α is given—suffice to solve numerically a wide range of geometrical problems. In analyzing how Ptolemy went about finding such numerical solutions we shall concentrate on the fundamental problems involving triangles; most geometrical problems can be reduced to problems involving only sides and angles of triangles, and these, as we shall now see, can be solved with Ptolemy's table of chords.

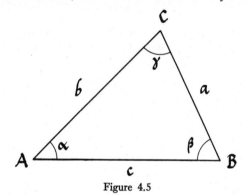

Figure 4.5

Let triangle ABC have sides a, b, c and angles α, β, γ (see Figure 4.5). We say that the triangle is *solved* when numerical values of these six quantities are found. First we observe that

$$\alpha + \beta + \gamma = 180°$$

so that if two angles are known, then so is the third.

We shall show that when values of three of these six quantities are given (clearly, no more than two may be angles), then the remaining three quantities can be computed if we have Ptolemy's table of chords at our disposal.

I may say here, to forestall misunderstanding, that Ptolemy does not discuss his trigonometric procedures in as general terms as I shall do below. What follows is distilled from numerous passages in the *Almagest*.

I. First, let us consider a *right* triangle, for this case is basic to all others.

We construct its circumscribed circle (see Figure 4.6) and note that hypotenuse AB is a diameter, so the centre of the circle is the midpoint

M of AB. The arc CB is then 2α, so a is the chord of 2α in a circle of diameter c. If the diameter were 120, the chord of 2α would be Ptolemy's crd 2α, so we get

$$\frac{\text{crd } 2\alpha}{120} = \frac{a}{c}.$$

Thus, if two of the three quantities α, a, and c are known we can find the third. Furthermore, from α we immediately find β, for

$$\beta = 90° - \alpha,$$

and from a and c we can find b by Pythagoras' theorem:

$$b = \sqrt{c^2 - a^2}.$$

So we see that with a table of chords, we can find all sides and angles in a right triangle if we know two of these quantities, provided one of them is a side.

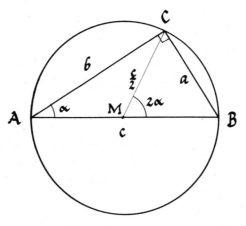

Figure 4.6

II. We shall now show that given an angle α and its adjacent sides b, c in a triangle, we can find the remaining side and angles.

Let us assume that α is acute. From C we drop the altitude h and denote its foot on c by H (see Figure 4.7). In the right triangle AHC we know the hypotenuse b and an angle α; thus we may find both h and $p = AH$ as the preceding example showed. If $p < c$, we find

$$q = c - p.$$

In the right triangle BHC we now know two sides, h and q, and can thus find both a and the angle β at B. The angle at C is now easily found since the sum of the three angles is 180°. The reader should consider the cases when $p \geq c$ and when α is obtuse.

Problems

4.1 If $p \geq c$, let $q = p - c$ and show how to determine q, a and β.

4.2 If α is obtuse, let $p = AH$, $q = p + c$ and show how to determine p, q, a and β.

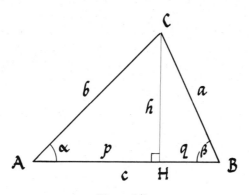

Figure 4.7

III. We next consider the case where two angles and a side, say b, are given. We can immediately find the third angle, for

$$\alpha + \beta + \gamma = 180°.$$

We assume that the triangle is acute. In the right triangle AHC (see Figure 4.7) we know b and α; we can therefore find p and h. In the right triangle CHB we now know h and β, so we can find a and q. Finally we get

$$c = p + q.$$

Problem

4.3 Give an analogous discussion in each of the cases: (a) α is obtuse, (b) β is obtuse, (c) γ is obtuse.

IV. We treat the case where an angle α, one side adjacent to α, and the side opposite α are given in a similar way by dividing the triangle into two right triangles. The reader should supply the details. (Note that there may be two different solutions to Problem 4.4.)

Problem

4.4 Given α and sides a and b, show how to obtain c, β and γ. Consider all possibilities.

V. The case where all three sides a, b, c are given is the most difficult. Ptolemy rarely has occasion to solve this problem, for in astronomical situations it is natural that angles are among the given quantities. But in *Almagest* VI, 17, in a passage dealing with eclipses, he does solve a problem of this sort, and he does it in the following manner.

Once again we divide the triangle into two right triangles by an altitude (see Figure 4.7), and we assume that H lies between A and B, i.e. that the altitude lies in the interior of the triangle (this will happen for at least one altitude). Our object is to find p and q, about which we know

$$p + q = c.$$

From triangle AHC we get

$$h^2 \;=\; b^2 - p^2$$

and from triangle BHC,

$$h^2 \;=\; a^2 - q^2.$$

Thus

$$b^2 - p^2 \;=\; a^2 - q^2$$

or

$$p^2 - q^2 \;=\; b^2 - a^2.$$

Here we assume that $b \geq a$; if $b < a$, we work with the equality $q^2 - p^2 = a^2 - b^2$. But

$$p^2 - q^2 \;=\; (p + q)(p - q) \;=\; c \cdot (p - q),$$

so

$$c \cdot (p - q) \;=\; b^2 - a^2$$

or

$$p - q \;=\; \frac{b^2 - a^2}{c}.$$

Since a, b, and c are given we may now compute $p - q$; and since $p + q$ was known, we can easily find both p and q. From the two right triangles we can then find α and β, and hence γ.

II–V exhaust all cases where an independent triple of quantities among a, b, c, α, β, and γ is given. We saw that the only basic formula in addition to the Pythagorean theorem was

$$\frac{a}{c} = \frac{\text{crd } 2\alpha}{120},$$

derived for a right triangle in I.

This means that whenever α is given, we must first double α, look up crd 2α, and divide this value by 120. The last operation is, however, quite simple, for to divide by 120 is the same as dividing by 2 and then by 60, and in the sexagesimal system we divide by 60 simply by moving the semicolon one place to the left. So what we have to do is, in essence, find half the chord of twice the arc. We have to do this os often that it would be reasonable to tabulate a new function, the half-chord of twice the arc, instead of the perhaps more natural chord function. This was, as a matter of fact, done by the Hindu astronomers (who were influenced both by Babylonian and early Greek theories).

The half-chord was called *jiva* in Sanskrit. The Muslim astronomers, in turn, learned both from the Greek tradition and from the Hindus, and they included in their works tables of half-chords, borrowing directly the Sanskrit name *jiva*. In Arabic, as in Hebrew, one often writes only the consonants of a word, leaving the vowels to be supplied by the reader. Now the unfamiliar Sanskrit word *jiva* has the same consonants as the common Arabic word *jaib*, which means *bay* or *pocket*. It was therefore not surprising that when Arabic astronomical works were translated into Latin, the translators who knew Arabic, but not Sanskrit, read the title of the tables of half-chords as *jaib* and turned it into the Latin word for *bay* or *pocket*, which is *sinus*. This is how our sine function received its curious name; and we have indeed already seen that half the chord of twice the arc is essentially what we now call sine of the arc.

Before we turn to the sequence of theorems by which Ptolemy shows how a table of chords can be constructed, it should be said that trigonometry surely did not originate with him. It is highly probable that Apollonius (*ca.* 200 B.C.) had this theory at his command, and it is certain that Hipparchos (*ca.* 150 B.C.) did. Menelaos (*ca.* 100 A.D.) perfected spherical trigonometry, and Ptolemy's treatment of this subject owes much to him. But it happens that Ptolemy's is the earliest (and one of the best) of surviving methodological treatments of the subject.

4.3   Ptolemy's Construction of the Table of Chords

We have seen how Ptolemy's table of chords looks, and what some of its uses are. I shall now set forth in detail how this table was constructed by presenting the material in the tenth chapter of the first book of the *Almagest*, following the same procedure I used in the chapters on Euclid and Archimedes: the theorems, their proofs, and order of presentation will be Ptolemy's; the language and notation will be mine.

First a point of notation: it will be convenient to let c_n mean the chord of $1/n$ of the circumference of a circle. If the radius is 60 we have

$$c_n = \operatorname{crd} \frac{360°}{n}.$$

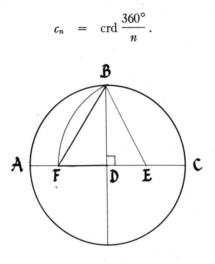

Figure 4.8

In a circle of center D (see Figure 4.8) we erect the perpendicular BD on the diameter AC. DC is bisected at E, and EF, equal to EB, is laid off along EA. Ptolemy now maintains that

$$\text{(i)} \quad DF = c_{10} \quad \text{and} \quad \text{(ii)} \quad BF = c_5.$$

We recognize the construction of the side of the decagon from Chapter 2, so (i) is clear. The solution to Problem 2.1 of Chapter 2 establishes relation (ii). Ptolemy refers to what must be Euclid XIII, 10 which states that

$$c_5^2 = c_6^2 + c_{10}^2.$$

This is the same as (ii), for c_6 is equal to the radius.

Ptolemy now proceeds to a calculation of c_{10} and c_5. With this norm, that the diameter is 120, we have

$$DE \ = \ 30, \quad DB \ = \ 60,$$

so

$$EB^2 \ = \ 30^2 + 60^2 \ = \ 4500, \quad EB \ = \ 67;4,55,$$

and hence

$$FD \ = \ FE - DE \ = \ EB - DE \ = \ 67;4,55 - 30 \ = \ 37;4,55 \ = \ c_{10}.$$

Thus

(1) $$c_{10} \ = \ \text{crd } 36° \ = \ 37;4,55.$$

Now $FD = 37;4,55$, so

$$FD^2 \ = \ 1375;4,15$$
$$DB^2 \ = \ 3600$$

and hence

$$FD^2 + DB^2 \ = \ BF^2 \ = \ 4975;4,15,$$

whence

$$BF \ = \ 70;32,3 \ = \ c_5.$$

Thus

(2) $$c_5 \ = \ \text{crd } 72° \ = \ 70;32,3.$$

We now have two entries for the table.

It is clear, since c_6 is the radius, that

(3) $$c_6 \ = \ \text{crd } 60° \ = \ 60.$$

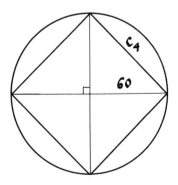

Figure 4.9

We have (see Figure 4.9) that

$$c_4^2 \;=\; 2 \cdot r^2 \;=\; 2 \cdot 60^2 \;=\; 7200$$

or

$$c_4 \;=\; 84;51,10.$$

Thus

(4) $$c_4 \;=\; \text{crd } 90° \;=\; 84;51,10.$$

This implies that Ptolemy's value for $\sqrt{2}$ is 1;24,51,10 which, incidentally, is the same as the one we found in the Old-Babylonian tablet in Chapter 1. Ptolemy takes for granted that one knows how to extract the square-root of a number, an operation performed very frequently. We do not know for certain which of several possible techniques he used.

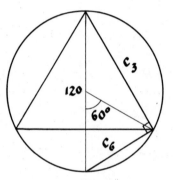

Figure 4.10

From Figure 4.10 we get

$$c_3^2 + c_6^2 \;=\; 4r^2$$

or

$$c_3^2 \;=\; 3r^2 \;=\; 10800, \qquad c_3 \;=\; 103;55,23.$$

Thus

(5) $$c_3 \;=\; \text{crd } 120° \;=\; 103;55,23.$$

This implies that Ptolemy's value for $\sqrt{3}$ is 1;43,55,23.

If we know the chord of an arc, we may find the chord of the supplementary arc as follows (see Figure 4.11):

$$\text{crd } (180° - \alpha) \;=\; \sqrt{(2r)^2 - \text{crd}^2\,\alpha} \;=\; \sqrt{14400 - \text{crd}^2\,\alpha}.$$

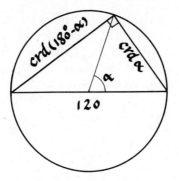

Figure 4.11

Ptolemy illustrates this by the example

$$\text{crd } 144° \;=\; \sqrt{14400 - \text{crd}^2\,36°} \;=\; \sqrt{14400 - 1375;4,15}$$

$$=\; \sqrt{13024;55,45} \;=\; 114;7,37.$$

Thus

(6) $$\text{crd } 144° \;=\; 114;7,37.$$

From these directly found chords we shall derive all the entries in the table by means of the following theorem:

If ABCD is an inscribable quadrilateral, then the sum of the products of opposite sides equals the product of the diagonals. In symbols (see Figure 4.12)

$$AB \cdot CD + BC \cdot AD \;=\; AC \cdot BD.$$

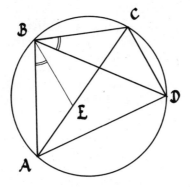

Figure 4.12

To prove this, we construct a point E on the diagonal AC so that angle ABE is equal to angle DBC. We first observe that triangle BCE and triangle BDA are similar, for the angles CBE and ABD are equal (by construction) and the angles BCA and BDA are equal (since they subtend the same arc). Hence

$$\frac{BC}{CE} = \frac{BD}{AD}$$

or

(i) $AD \cdot BC = CE \cdot BD.$

Next we note that triangle BAE and triangle BDC are similar, for the double-marked angles at B are equal, and so are angles BAC and BDC (since they subtend the same arc). Hence

$$\frac{AB}{BD} = \frac{AE}{CD}$$

or

(ii) $AB \cdot CD = AE \cdot BD.$

Adding (i) and (ii) we obtain

$$AD \cdot BC + AB \cdot CD = CE \cdot BD + AE \cdot BD$$
$$= BD \cdot (CE + AE)$$
$$= BD \cdot AC,$$

which was to be proved.

This theorem is usually called Ptolemy's theorem, though it was certainly discovered long before his time. It is often proved in elementary mathematics courses, but it looks curiously unmotivated when removed from its natural setting. We shall now see what its proper purpose is.

Ptolemy now proves that, given two arcs and their chords, we may find the chord of the difference of these arcs in terms of the chords of the given arcs.

In Figure 4.13 we are given AB and AC and are to show that we can find BC. We draw the diameter AD. We can find chords BD and CD of the supplementary arcs as shown above, and Ptolemy's theorem gives

$$AB \cdot CD + BC \cdot AD = AC \cdot BD.$$

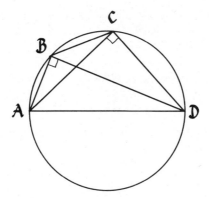

Figure 4.13

Since $AD = 120$,

$$120 \cdot BC \quad = \quad AC \cdot BD - AB \cdot CD.$$

Everything on the right side is known, and so we can find BC. Calling the arcs α and β we may write this result as

(7) $120 \, \text{crd} \, (\beta - \alpha) \quad = \quad \text{crd} \, \beta \, \text{crd} \, (180° - \alpha) - \text{crd} \, \alpha \, \text{crd} \, (180° - \beta),$

which is strongly reminiscent of the formula

$$\sin (\beta - \alpha) \quad = \quad \sin \beta \cos \alpha - \sin \alpha \cos \beta.$$

Ptolemy now says that from crd 72° and crd 60° we may find

$$\text{crd} \, 12° \quad = \quad \text{crd} \, (72° - 60°).$$

Next Ptolemy shows how, given the chord of an arc, we may find the chord of half the given arc.

In Figure 4.14, let BC be the given chord. Draw the diameter AC through C, and let D be the midpoint of the arc BC. Drop the perpendicular DF from D on AC.

We are to find DC. To this end we shall first show that FC, the projection of DC on AC, is equal to $\frac{1}{2}(AC - AB)$. To do this, we mark off $AE = AB$; we have then that triangle BAD and triangle EAD are congruent, for the angles at A are equal (D bisects the arc BC), and so are two pairs of adjacent sides. Hence

$$DE \quad = \quad BD;$$

but

$$BD \quad = \quad DC, \quad \text{so} \quad DE \quad = \quad DC,$$

i.e. triangle EDC is isosceles. Thus DF, an altitude of an isosceles triangle, is a median, so that

$$EF = FC.$$

This establishes the desired result, for

$$FC = \tfrac{1}{2}(AC - AE) = \tfrac{1}{2}(AC - AB).$$

Now $AC = 120$, and AB can be found as the chord of the supplementary arc to arc BC, so FC is computable.

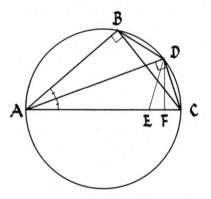

Figure 4.14

To show that DC can be found, we apply to triangle ACD the theorem that a side in a right triangle is the mean proportional between its projection on the hypotenuse and the entire hypotenuse, and obtain

$$DC^2 = AC \cdot FC = 120 \cdot FC = 120 \cdot \tfrac{1}{2}(AC - AB).$$

If we call the given arc α, we may write the result in the form

(8)
$$\operatorname{crd}^2 \frac{\alpha}{2} = 120 \cdot \frac{1}{2}\left(120 - \operatorname{crd}\left(180° - \alpha\right)\right)$$

$$= 60 \cdot \left(120 - \operatorname{crd}\left(180° - \alpha\right)\right),$$

which is reminiscent of the half-angle formula

$$\sin^2 \frac{\alpha}{2} = \frac{1}{2}\left(1 - \cos\alpha\right).$$

Ptolemy now points out that, starting with crd 12°, one can find crd 6°, crd 3°, crd $1\tfrac{1}{2}$°, and crd $\tfrac{3}{4}$°, by using this procedure repeatedly.

This yields

$$\text{crd } 1\tfrac{1}{2}° \quad = \quad 1;34,15$$

$$\text{crd } \tfrac{3}{4}° \quad = \quad 0;47,8.$$

Having found chords for such small angles, we could now make a systematic table of chords with, for example, entries differing by $1\tfrac{1}{2}°$, if only we had an addition formula for chords. This is precisely what Ptolemy supplies next; he shows that, given the chords of two arcs, we may find the chord of the sum of the arcs.

Let AB and BC be the given chords (see Figure 4.15); we shall find AC. We draw the diameters AD and BE and join points as in the figure.

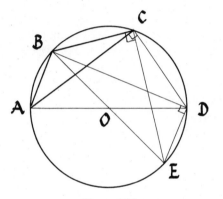

Figure 4.15

We can find BD and CE as chords of the supplements of the given arcs, and we note that $DE = AB$. We apply Ptolemy's theorem to the quadrilateral $BCDE$:

$$BD \cdot CE \quad = \quad BC \cdot DE + BE \cdot CD,$$

or

$$BE \cdot CD \quad = \quad BD \cdot CE - BC \cdot DE.$$

But $BE = 120$, and $DE = AB$, so

$$120 \cdot CD \quad = \quad BD \cdot CE - BC \cdot AB.$$

Since everything on the right is known, we can find CD and hence AC.

Calling the given arcs α and β, we may write this result in the form

(9) $120 \cdot \text{crd}\left(180° - (\alpha + \beta)\right)$

$$= \quad \text{crd}\left(180° - \alpha\right) \cdot \text{crd}\left(180° - \beta\right) - \text{crd } \alpha \text{ crd } \beta,$$

which is the counterpart of our addition formula

$$\cos (\alpha + \beta) \;=\; \cos \alpha \cos \beta - \sin \alpha \sin \beta.$$

We have now seen the real content of Ptolemy's theorem. It is not an ingenious but quaint result on inscribable quadrilaterals, but it is precisely the Greek equivalent of our addition formulas for trigonometric functions.

We are now in a position to find

$$\operatorname{crd} 4\tfrac{1}{2}° \;=\; \operatorname{crd} (3° + 1\tfrac{1}{2}°),$$

$$\operatorname{crd} 6° \;=\; \operatorname{crd} (4\tfrac{1}{2}° + 1\tfrac{1}{2}°),$$

and so on. It is, however, Ptolemy's desire to have the entries in his table increase by steps of $\tfrac{1}{2}°$; so we need to fill the gaps in the sequence $1\tfrac{1}{2}°$, $3°$, $4\tfrac{1}{2}°$, $6°$, $7\tfrac{1}{2}°$, $\cdots$. If we knew $\operatorname{crd} \tfrac{1}{2}°$, we could do this by means of the addition and subtraction formulas.

In order to find $\operatorname{crd} \tfrac{1}{2}°$ from the known $\operatorname{crd} 1\tfrac{1}{2}°$ in a manner similar to Ptolemy's previous procedures, we should need a relation between $\operatorname{crd} \alpha$ and $\operatorname{crd} 3\alpha$, for $1\tfrac{1}{2}°$ is thrice $\tfrac{1}{2}°$. Such a relation is provided by the modern identity

$$\sin 3\alpha \;=\; 3 \sin \alpha - 4 \sin^3 \alpha$$

or, transformed to chords with Ptolemy's norm,

$$\frac{1}{120} \cdot \operatorname{crd} 6\alpha \;=\; \frac{3}{120} \cdot \operatorname{crd} 2\alpha - \frac{4}{120^3} \cdot \operatorname{crd}^3 2\alpha,$$

or

(i) $$\operatorname{crd} 3\beta \;=\; 3 \operatorname{crd} \beta - \frac{1}{60^2} \cdot \operatorname{crd}^3 \beta.$$

So if $\operatorname{crd} 3\beta$ is given, we must solve a cubic equation in order to find $\operatorname{crd} \beta$. This was, in fact, done much later when a Persian astronomer al-Kāshī (d. 1429) found $\sin 1°$ from $\sin 3°$ by an ingenious method of his own device for solving the trisection equation as accurately as desired, but Ptolemy seeks a different way out of the dilemma. He does not derive the identity (i), nor anything equivalent to it, in the *Almagest*. It is, however, perfectly clear that he explored this approach as a possibility, for when explaining the necessity of finding $\operatorname{crd} \tfrac{1}{2}°$ he continues: "However, when for example the chord of $1\tfrac{1}{2}°$ is given, one cannot find the chord of one-third of this arc by geometrical construction."

The expression "by geometrical construction" is, in the Greek, literally, "by means of lines (or curves)". The precise meaning of this phrase is not completely clear, although we have just seen the nature of the trouble Ptolemy is facing. It may mean that the trisection of an angle cannot be executed by compasses and straightedge; and this, of course, is true and would be of importance here, for these constructions can readily be paralleled by computations involving nothing worse than square-root extractions (this can be shown fairly simply). But an angle can certainly be trisected by geometrical constructions employing other means, as we saw in the chapter on Archimedes.

The assertion that the problem cannot be solved "by means of lines", may also be interpreted to mean that it is not "plane", which in this connection would mean that it leads to an equation of more than second degree (a "cube" is not a plane figure, but a "square" is). This interpretation seems to me more plausible, for Hipparchos uses the identical term "by means of lines" when he tells, in one of his few surviving works, that he can solve spherical problems in the plane (using stereographic projection).

Ptolemy cannot solve a cubic equation, but fortunately he does not need to in order to find crd $\frac{1}{2}°$ to two correct sexagesimals.

It was found above that

$$\text{crd } 1\tfrac{1}{2}° \;=\; 1;34,15, \qquad \text{crd } \tfrac{3}{4}° \;=\; 0;47,8,$$

i.e. that crd $\frac{3}{4}°$ is approximately half of crd $1\frac{1}{2}°$. Since $\frac{3}{4}°$ is half of $1\frac{1}{2}°$, it may not be unreasonable to guess that crd $1°$ is $\frac{2}{3}$ of crd $1\frac{1}{2}°$, even as $1°$ is $\frac{2}{3}$ of $1\frac{1}{2}°$. This would give

$$\text{crd } 1° \;=\; \tfrac{2}{3}\cdot 1;34,15 \;=\; 1;2,50.$$

Problem

4.5 Show that the approximation

$$\text{crd } \frac{\beta}{2} \;=\; \frac{1}{2}\,\text{crd } \beta,$$

accurate to two sexagesimal places for $\beta = 1\frac{1}{2}°$, is very inaccurate for large β. For example, compute $\frac{1}{2}c_6 = \frac{1}{2}\,\text{crd } 60°$ and compare it to

$$c_{12} \;=\; \text{crd } 30° \;=\; \text{crd } \frac{60°}{2}$$

as found by the half-angle formula (8).

The value for chord 1° that we find in Ptolemy's table is, in fact, 1;2,50. Ptolemy may have arrived at this value by conjecture, just as we did, but he is not satisfied with reasonable guesses. He requires that the accuracy of the value to two sexagesimals be established beyond any doubt, and to accomplish this he proves the following theorem.

If we are given two unequal chords, chord α greater than chord β, then

$$\frac{\text{crd } \alpha}{\text{crd } \beta} < \frac{\alpha}{\beta}.$$

(As before, we only consider arcs $<180°$.)

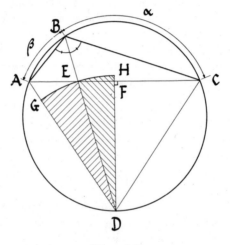

Figure 4.16

In Figure 4.16, the two arcs are the ones determined by the chords AB and BC, where $AB < BC$. We wish to prove that

$$\frac{BC}{AB} < \frac{\text{arc } BC}{\text{arc } AB}.$$

First we bisect the angle at B, and extend the angle bisector BE (E on AC) until it meets the circle at D. We then have

$$AD = DC$$

because they are subtended by equal angles.

There is a theorem stating that the bisector of an angle in a triangle cuts the opposite side into segments proportional to the adjacent sides

(Euclid VI, 3). Applied to triangle ABC it yields

(i)
$$\frac{AE}{EC} = \frac{AB}{BC}$$

which we shall use later. What we need now is that

$$AE \ < \ EC,$$

and this follows from (i) since

$$AB \ < \ BC. \ \cdot$$

From D we drop the perpendicular DF on AC; F is the midpoint of AC. We now have

$$AD \ > \ ED \ > \ FD,$$

so a circle of centre D and radius ED will cut AD between A and D at G and DF (extended beyond F) at H. Therefore, we see, when considering the two shaded circular sectors, that

$$\text{sector } DEH \ > \ \text{triangle } DEF$$

and

$$\text{sector } DEG \ < \ \text{triangle } DEA.$$

Thus

(ii)
$$\frac{\text{triangle } DEF}{\text{triangle } DEA} \ < \ \frac{\text{sector } DEH}{\text{sector } DEG}.$$

Now the two triangles have the altitude DF in common, so the ratio of their areas is the same as the ratio of their bases. The left side of the inequality (ii) can therefore be replaced by EF/EA. Moreover, the areas of sectors of a circle have the same ratio as the corresponding central angles, so the right side of (ii) can be replaced by $\angle EDH/\angle EDG$. Thus we have

$$\frac{EF}{EA} \ < \ \frac{\angle EDH}{\angle EDG}.$$

From this it follows (by adding 1 to both sides of the inequality) that

$$\frac{EF + EA}{EA} \ < \ \frac{\angle EDH + \angle EDG}{\angle EDG}$$

or

$$\frac{AF}{EA} \ < \ \frac{\angle GDH}{\angle EDG}.$$

Hence

$$\frac{2AF}{EA} < \frac{2\sphericalangle GDH}{\sphericalangle EDG}$$

or

$$\frac{AC}{EA} < \frac{\sphericalangle ADC}{\sphericalangle EDG}.$$

Subtracting 1 from both sides of this inequality yields

$$\frac{AC - EA}{EA} < \frac{\sphericalangle ADC - \sphericalangle EDG}{\sphericalangle EDG}$$

or

(iii) $$\frac{EC}{EA} < \frac{\sphericalangle CDE}{\sphericalangle EDG}.$$

Using (i) and the fact that an angle in a circle is half the arc it subtends, we can write (iii) in the form

$$\frac{BC}{AB} < \frac{\text{arc } BC}{\text{arc } AB}.$$

This completes the proof.

Ptolemy now applies this theorem to two cases:

1. $\alpha = 1\frac{1}{2}°$, $\beta = 1°$;   2. $\alpha = 1°$, $\beta = \frac{3}{4}°$.

In case 1 he gets

$$\frac{\text{crd } 1\frac{1}{2}°}{\text{crd } 1°} < \frac{1\frac{1}{2}}{1} = \frac{3}{2},$$

so

$$\text{crd } 1° > \tfrac{2}{3} \cdot \text{crd } 1\frac{1}{2}° = \tfrac{2}{3} \cdot 1;34,15 = 1;2,50.$$

In case 2 he gets

$$\frac{\text{crd } 1°}{\text{crd } \frac{3}{4}°} < \frac{1}{(\frac{3}{4})} = \frac{4}{3},$$

so

$$\text{crd } 1° < \tfrac{4}{3} \cdot \text{crd } \tfrac{3}{4}° = \tfrac{4}{3} \cdot 0;47,8 = 1;2,50.$$

Ptolemy reasons that since crd $1°$ is both larger than and smaller than 1;2,50 it must be equal to 1;2,50 (he means, of course, to two sexagesimals) and from this he obtains, by (8),

$$\text{crd } \tfrac{1}{2}° \;=\; 0;31,25.$$

He can now complete his table in steps of $\tfrac{1}{2}°$.

Problems

4.6 Prove the theorem (Euclid VI, 3) used on page 122, that the bisector AE of $\sphericalangle A$ in triangle ABC cuts the side BC into segments BE and EC such that $BE/EC = AB/AC$. (Hint: Extend AB beyond A to D so that $AD = AC$ and join DC.)

4.7 From Ptolemy's value for crd $1°$, i.e. crd $1° = 1;2,50$, find the perimeter of an inscribed 360-gon, and hence a sexagesimal approximation to π. Convert this number to a decimal and compare it to our value of π and to Archimedes' approximation

$$3\tfrac{10}{71} \;<\; \pi \;<\; 3\tfrac{10}{70}.$$

Is Ptolemy's value for crd $1°$ too large or too small?

Incidentally, al-Kāshī (d.1429)—whom I mentioned above for his solution of the trisection equation—finds the following approximations:†

$$2\pi \sim 6;16,59,28,1,34,51,46,14,50$$

$$\sim 6.2831853071795865.$$

These are correct as far as they go. He gives both the sexagesimal and the decimal representation, for he claims decimal fractions as an invention of his own. [Simon Stevin (1548–1620) is credited with the introduction of decimal fractions in the West.]

Al-Kāshī achieves this astounding computational feat by finding the lengths of an inscribed and a circumscribed regular polygon of $3 \cdot 2^{28} = 805,306,368$ sides, beginning with the side of a regular triangle and applying the half-arc formula 28 times.

It should be clear to anyone who has read this chapter that the commonly held notion that Greek mathematics is entirely geometrical is not quite correct. Greek mathematicians were perfectly capable of doing numerical work when they had to; indeed, one has to look far and wide to find Ptolemy's equal as a computer.

---

† P. Luckey, *Der Lehrbrief über den Kreisumfang ... von ... al-Kāšī*, Abh. d. deutschen Akad. d. Wiss. zu Berlin, 1953.

I regret that it was not possible to present some of Ptolemy's original work, but its proper appreciation requires more knowledge of both mathematics and astronomy than I have wanted to presuppose in my readers. Still, Ptolemy's faculty for lucid and orderly presentation may shine through the preceding sequence of theorems.

APPENDIX

Ptolemy's Epicyclic Models

Figure 4.17 represents the epicyclic model which Ptolemy used for any planet except Mercury. Mercury has always been a bothersome planet and required special attention even then. I shall describe this model in the greatest brevity, for it may be of interest to those who have some knowledge of astronomy; but I shall neither analyze Ptolemy's procedure leading to this model, nor attempt to justify its great efficiency by comparing it with present-day theories.

In Figure 4.17 the paper is the plane of the ecliptic viewed from its north pole. The *ecliptic* or *zodiac* is the apparent path among the fixed stars of the sun during its yearly travel; it is a great circle on the celestial sphere and was divided into twelve parts or zodiacal signs (Aries, Taurus, etc.) of 30° each. The zodiac was used as a circle of reference for describing the wandering of the planets among the fixed stars, for a planet is always seen near the ecliptic. In the figure we simplify the situation and assume that the planet is in the plane of the ecliptic, as Ptolemy does at first.

The Planet P now moves on the epicycle, whose centre C' moves on the deferent. The observer sits at O outside the center C of the deferent. The distance $OC = e$ is called the *eccentricity* of the deferent, and the line through O and C is fixed relative to the fixed stars. The longitude of the apogee of the deferent (see the figure) is constant (except for precession). Ptolemy normalized the model so that the radius of the deferent is 1; the radius of the epicycle is r in this unit.

The motion of P is governed in the following way: C' moves uniformly counter-clockwise, not about C, but about a point E called the *equant point*; E is symmetrical to O with respect to C, i.e., $OE = e$. This means that the angle α grows by the same amount every day. P now moves on the epicycle so that the angle β grows by a constant amount every day. Thus, if the date is known, so are α and β, provided we know their values at some particular fixed date (the epoch), and their rates of growth.

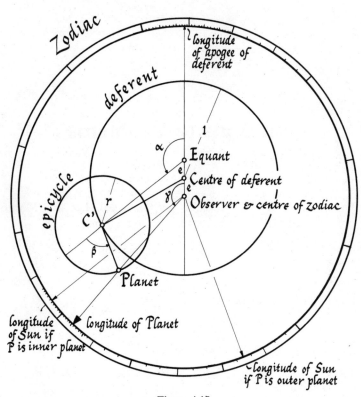

Figure 4.17

Incidentally, these angles vary in such a way that the direction from O to the (mean) sun is the same as the direction from E to C' for Venus, and from C' to P for outer planets.

Now, if we know the parameters of this model,

1. r, the radius of the epicycle
2. e, the eccentricity of the deferent
3. the longitude of the deferent's apogee
4, 5. the values of α and β at epoch
6, 7. the rate of increase of α and β,

then we may find the longitude of the planet at any given time. For given the time we can find α and β, and the figure shows that then we may construct the angle γ which, when added to the longitude of the apogee of the deferent, gives the longitude of the planet.

If our aim is to find the longitude of the planet as a number of degrees, we must have methods for carrying out this construction numerically. Such methods are precisely what I have set forth above.

Solutions to Problems

1.1 The semicolon should be in this position: 0;0,44,26,40. Multiplication by 81 verifies the assertion of the problem.

1.4 $1/n$ $(1 < n \leq 20)$ has a finite binary expansion if

$$n = 2, 4, 8, 16.$$

$1/n$ has a finite decimal expansion if

$$n = 2, 4, 5, 8, 10, 16, 20.$$

$1/n$ has a finite sexagesimal expansion if

$$n = 2, 3, 4, 5, 6, 8, 9, 10, 12, 15, 16, 18, 20.$$

1.7 Since p, q are relatively prime and not both odd, one is even, the other odd; hence $p^2 + q^2$ and $p^2 - q^2$ are odd. Therefore $x = 100 = 2pq$ is the *even* member of the triple. Thus $pq = 2 \cdot 5^2$, and $p = 25, q = 2$ or $p = 50, q = 1$ are the only possible p, q satisfying the theorem. These yield the triples

$$100, \quad 621, \quad 629 \quad \text{and} \quad 100, \quad 2499, \quad 2501$$

respectively.

By analogous reasoning, we find that $2pq = 210$, or $pq = 105 = 3 \cdot 5 \cdot 7$ yields no solutions since neither p nor q can be even, and that $2pq = 420$, or $pq = 210 = 2 \cdot 3 \cdot 5 \cdot 7$ yields eight solutions, corresponding to

$$q = 1, 2, 3, 5, 7, 2 \cdot 3, 2 \cdot 5, 3 \cdot 5.$$

If $x = 35$, x must be an *odd* member of the triple. It cannot be $p^2 + q^2$ because 35 is not the sum of two squares, so it must be $p^2 - q^2$. Hence

$$p^2 - q^2 = (p + q)(p - q) = 35,$$

128

and the only solutions satisfying the theorem are

$$p = 6, \qquad q = 1; \qquad p = 18, \qquad q = 17.$$

1.8 The missing part of the multiplication table appears below.

1	59	58	57	56	55	54	53	52	51	50	49	48	47	46	45	44	43	42	41
2	1,58	1,56	1,54	1,52	1,50	1,48	1,46	1,44	1,42	1,40	1,38	1,36	1,34	1,32	1,30	1,28	1,26	1,24	1,22
3	2,57	2,54	2,51	2,48	2,45	2,42	2,39	2,36	2,33	2,30	2,27	2,24	2,21	2,18	2,15	2,12	2,9	2,6	2,3
4	3,56	3,52	3,48	3,44	3,40	3,36	3,32	3,28	3,24	3,20	3,16	3,12	3,8	3,4	3,0	2,56	2,52	2,48	2,44
5	4,55	4,50	4,45	4,40	4,35	4,30	4,25	4,20	4,15	4,10	4,5	4,0	3,55	3,50	3,45	3,40	3,35	3,30	3,25
6	5,54	5,48	5,42	5,36	5,30	5,24	5,18	5,12	5,6	5,0	4,54	4,48	4,42	4,36	4,30	4,24	4,18	4,12	4,6
7	6,53	6,46	6,39	6,32	6,25	6,18	6,11	6,4	5,57	5,50	5,43	5,36	5,29	5,22	5,15	5,8	5,1	4,54	4,47
8	7,52	7,44	7,36	7,28	7,20	7,12	7,4	6,56	6,48	6,40	6,32	6,24	6,16	6,8	6,0	5,52	5,44	5,36	5,28
9	8,51	8,42	8,33	8,24	8,15	8,6	7,57	7,48	7,39	7,30	7,21	7,12	7,3	6,54	6,45	6,36	6,27	6,18	6,9
10	9,50	9,40	9,30	9,20	9,10	9,0	8,50	8,40	8,30	8,20	8,10	8,0	7,50	7,40	7,30	7,20	7,10	7,0	6,50
11	10,49	10,38	10,27	10,16	10,5	9,54	9,43	9,32	9,21	9,10	8,59	8,48	8,37	8,26	8,15	8,4	7,53	7,42	7,31
12	11,48	11,36	11,24	11,12	11,0	10,48	10,36	10,24	10,12	10,0	9,48	9,36	9,24	9,12	9,0	8,48	8,36	8,24	8,12
13	12,47	12,34	12,21	12,8	11,55	11,42	11,29	11,16	11,3	10,50	10,37	10,24	10,11	9,58	9,45	9,32	9,19	9,6	8,53
14	13,46	13,32	13,18	13,4	12,50	12,36	12,22	12,8	11,54	11,40	11,26	11,12	10,58	10,44	10,30	10,16	10,2	9,48	9,34
15	14,45	14,30	14,15	14,0	13,45	13,30	13,15	13,0	12,45	12,30	12,15	12,0	11,45	11,30	11,15	11,0	10,45	10,30	10,15
16	15,44	15,28	15,12	14,56	14,40	14,24	14,8	13,52	13,36	13,20	13,4	12,48	12,32	12,16	12,0	11,44	11,28	11,12	10,56
17	16,43	16,26	16,9	15,52	15,35	15,18	15,1	14,44	14,27	14,10	13,53	13,36	13,19	13,2	12,45	12,28	12,11	11,54	11,37
18	17,42	17,24	17,6	16,48	16,30	16,12	15,54	15,36	15,18	15,0	14,42	14,24	14,6	13,48	13,30	13,12	12,54	12,36	12,18
19	18,41	18,22	18,3	17,44	17,25	17,6	16,47	16,28	16,9	15,50	15,31	15,12	14,53	14,34	14,15	13,56	13,37	13,18	12,59
20	19,40	19,20	19,0	18,40	18,20	18,0	17,40	17,20	17,0	16,40	16,20	16,0	15,40	15,20	15,0	14,40	14,20	14,0	13,40
21	20,39	20,18	19,57	19,36	19,15	18,54	18,33	18,12	17,51	17,30	17,9	16,48	16,27	16,6	15,45	15,24	15,3	14,42	14,21
22	21,38	21,16	20,54	20,32	20,10	19,48	19,26	19,4	18,42	18,20	17,58	17,36	17,14	16,52	16,30	16,8	15,46	15,24	15,2
23	22,37	22,14	21,51	21,28	21,5	20,42	20,19	19,56	19,33	19,10	18,47	18,24	18,1	17,38	17,15	16,52	16,29	16,6	15,43
24	23,36	23,12	22,48	22,24	22,0	21,36	21,12	20,48	20,24	20,0	19,36	19,12	18,48	18,24	18,0	17,36	17,12	16,48	16,24
25	24,35	24,10	23,45	23,20	22,55	22,30	22,5	21,40	21,15	20,50	20,25	20,0	19,35	19,10	18,45	18,20	17,55	17,30	17,5
26	25,34	25,8	24,42	24,16	23,50	23,24	22,58	22,32	22,6	21,40	21,14	20,48	20,22	19,56	19,30	19,4	18,38	18,12	17,46
27	26,33	26,6	25,39	25,12	24,45	24,18	23,51	23,24	22,57	22,30	22,3	21,36	21,9	20,42	20,15	19,48	19,21	18,54	18,27
28	27,32	27,4	26,36	26,8	25,40	25,12	24,44	24,16	23,48	23,20	22,52	22,24	21,56	21,28	21,0	20,32	20,4	19,36	19,8
29	28,31	28,2	27,33	27,4	26,35	26,6	25,37	25,8	24,39	24,10	23,41	23,12	22,43	22,14	21,45	21,16	20,47	20,18	19,49
30	29,30	29,0	28,30	28,0	27,30	27,0	26,30	26,0	25,30	25,0	24,30	24,0	23,30	23,0	22,30	22,0	21,30	21,0	20,30
31	30,29	29,58	29,27	28,56	28,25	27,54	27,23	26,52	26,21	25,50	25,19	24,48	24,17	23,46	23,15	22,44	22,13	21,42	21,11
32	31,28	30,56	30,24	29,52	29,20	28,48	28,16	27,44	27,12	26,40	26,8	25,36	25,4	24,32	24,0	23,28	22,56	22,24	21,52
33	32,27	31,54	31,21	30,48	30,15	29,42	29,9	28,36	28,3	27,30	26,57	26,24	25,51	25,18	24,45	24,12	23,39	23,6	22,33
34	33,26	32,52	32,18	31,44	31,10	30,36	30,2	29,28	28,54	28,20	27,46	27,12	26,38	26,4	25,30	24,56	24,22	23,48	23,14
35	34,25	33,50	33,15	32,40	32,5	31,30	30,55	30,20	29,45	29,10	28,35	28,0	27,25	26,50	26,15	25,40	25,5	24,30	23,55
36	35,24	34,48	34,12	33,36	33,0	32,24	31,48	31,12	30,36	30,0	29,24	28,48	28,12	27,36	27,0	26,24	25,48	25,12	24,36
37	36,23	35,46	35,9	34,32	33,55	33,18	32,41	32,4	31,27	30,50	30,13	29,36	28,59	28,22	27,45	27,8	26,31	25,54	25,17
38	37,22	36,44	36,6	35,28	34,50	34,12	33,34	32,56	32,18	31,40	31,2	30,24	29,46	29,8	28,30	27,52	27,14	26,36	25,58
39	38,21	37,42	37,3	36,24	35,45	35,6	34,27	33,48	33,9	32,30	31,51	31,12	30,33	29,54	29,15	28,36	27,57	27,18	26,39
40	39,20	38,40	38,0	37,20	36,40	36,0	35,20	34,40	34,0	33,20	32,40	32,0	31,20	30,40	30,0	29,20	28,40	28,0	27,20
41	40,19	39,38	38,57	38,16	37,35	36,54	36,13	35,32	34,51	34,10	33,29	32,48	32,7	31,26	30,45	30,4	29,23	28,42	28,1
42	41,18	40,36	39,54	39,12	38,30	37,48	37,6	36,24	35,42	35,0	34,18	33,36	32,54	32,12	31,30	30,48	30,6	29,24	
43	42,17	41,34	40,51	40,8	39,25	38,42	37,59	37,16	36,33	35,50	35,7	34,24	33,41	32,58	32,15	31,32	30,49		
44	43,16	42,32	41,48	41,4	40,20	39,36	38,52	38,8	37,24	36,40	35,56	35,12	34,28	33,44	33,0	32,16			
45	44,15	43,30	42,45	42,0	41,15	40,30	39,45	39,0	38,15	37,30	36,45	36,0	35,15	34,30	33,45				
46	45,14	44,28	43,42	42,36	42,10	41,24	40,38	39,52	39,6	38,20	37,34	36,48	36,2	35,16					
47	46,13	45,26	44,39	43,52	43,5	42,18	41,31	40,44	39,57	39,10	38,23	37,36	36,49						
48	47,12	46,24	45,36	44,48	44,0	43,12	42,24	41,36	40,48	40,0	39,12	38,24							
49	48,11	47,22	46,33	45,44	44,55	44,6	43,17	42,28	41,39	40,50	40,1								
50	49,10	48,20	47,30	46,40	45,50	45,0	44,10	43,20	42,30	41,40									

2.1 We first introduce the notation c_n for the chord corresponding to $1/n$ of the circle's circumference or, what is the same, the side of a regular n-gon. Thus, in Figures 2.5 and 2.6,

$$x = c_{10}.$$

We wish to prove that CE in Figure 2.6 is c_5, or that

$$c_5^2 = c_{10}^2 + r^2.$$

In Figure 2.5, drop the altitude AD from A of $\triangle ACB$. Since $x = c_{10}$ and since AD is half the chord corresponding to twice the arc AB, we have

$$AD = \tfrac{1}{2}c_5.$$

We note that $OC = x = c_{10}$ so that $CB = r - c_{10}$, and
$$DB = \tfrac{1}{2}(r - c_{10}).$$

In the right $\triangle ADB$ we have
$$AD^2 + DB^2 = AB^2$$
or
$$\tfrac{1}{4}c_5^2 + \tfrac{1}{4}(r - c_{10})^2 = c_{10}^2$$

which reduces to
$$c_5^2 = 3c_{10}^2 + 2rc_{10} - r^2.$$

However, from equation (1) on p. 55 we obtain, after slight reduction,
$$rc_{10} = r^2 - c_{10}^2$$

which, substituted in the preceding equation, yields
$$c_5^2 = c_{10}^2 + r^2, \qquad \text{q.e.d.}$$

3.1 In Figure 3.4, draw the semi-circle with diameter $CD = 2a$. From its centre E, draw its radius to A. In isosceles $\triangle AOE$, the base angles are $\alpha - \beta$; in isosceles $\triangle EAD$, the base angles are β. But $\sphericalangle OEA = \alpha - \beta$ is exterior to $\triangle EAD$, so
$$\alpha - \beta = 2\beta.$$

It follows that $\alpha = 3\beta$.

4.7 Since
$$\text{crd } 1° \sim 1;2,50$$

the perimeter of a 360-gon is
$$360 \text{ crd } 1° \sim 6,17 \sim \text{circumference of circle}.$$

Therefore
$$\pi = \frac{\text{circumference}}{\text{diameter}} \sim \frac{6,17}{2,0} = 3;8,30$$

or, in decimals, $3.141666\cdots$. Our modern decimal expansion of π begins with
$$3.14159\cdots,$$

and Archimedes' lower and upper bounds are
$$3.14084\cdots < \pi < 3.14285\cdots.$$

Ptolemy's value for crd $1°$ is slightly too large.

Suggestions for Further Reading

1. Otto Neugebauer, *The Exact Sciences in Antiquity*. 2nd edition. Providence: Brown University Press, 1957. (Also available in a paperback edition as a Harper Torchbook.)

   An outstanding treatment of a variety of topics from mathematics and astronomy in antiquity by the leading scholar in the field; contains also a rich bibliography.

2. B. L. van der Waerden, *Science Awakening*. 2nd edition. Oxford University Press, 1961.

   An excellent and methodical presentation of ancient mathematics by a scholar who is also well-known for his important contributions to modern mathematics. A second volume, on ancient astronomy, is in preparation.

These two books have relevance for all four chapters. For the chapters on Greek mathematics one can further consult:

3. Sir Thomas Heath, *A History of Greek Mathematics*. 2 vols. Oxford, 1921.

   A standard work of great completeness on Greek mathematics. Obsolete as far as pre-Greek mathematics is concerned. It exists in a condensed version as:

4. Sir Thomas L. Heath, *A Manual of Greek Mathematics*. 1 vol. Oxford, 1931. Republished 1963 by Dover Publications.

For the separate chapters the following references are readily available:

Chapter I

5. Otto Neugebauer and A. Sachs, *Mathematical Cuneiform Texts*. New Haven, 1945.

   A scholarly edition of mathematical texts, mostly from American collections.

For an elementary and detailed proof of the theorem on Pythagorean numbers see:

6. Hans Rademacher and Otto Toeplitz, *The Enjoyment of Mathematics*. Princeton: Princeton University Press, 1957.

   An elementary book written with great competence and charm.

An excellent introduction to archeological problems and techniques in Mesopotamia is found in:

7. Edward Chiera, *They Wrote on Clay*. Phoenix Books (paperback). Several editions since 1938.

Chapter II

8. *The Thirteen Books of Euclid's Elements*, translated from the text of Heiberg, with introduction and commentary, by Sir Thomas L. Heath. 3 vols. New York: Dover Publications, 1956.

   The best English version of Euclid, with extensive notes and comments.

Chapter III

Archimedes' works are translated, and transcribed into modern mathematical notation in:

9. *The Works of Archimedes*, ed. T. L. Heath. New York: Dover Publications.

For a profound analysis of Archimedes' works see:

10. E. J. Dijksterhuis, *Archimedes*. Copenhagen: E. Munksgaard, 1956.

For discussions of constructibility (none is elementary) the following books are available in paperbacks:

11. *Monographs on Topics of Modern Mathematics*, ed. J. W. A. Young. New York: Dover Publications, 1955 (chapter by L. E. Dickson).

12. Felix Klein, *Famous Problems of Elementary Geometry*. New York: Dover Publications, 1956.

See also:

13. R. Courant and H. Robbins, *What is Mathematics?* Oxford, 1941.

    A masterpiece of elementary mathematical exposition, and a book of surprising scope and richness.

For a brief survey of the history of mathematics from antiquity to modern times the reader may turn to:

14. Dirk J. Struik, *A Concise History of Mathematics*. New York: Dover Publications, 1948.

Aaboe, Asger   QA
  Episodes from the   22
early history of mathe-   A11
matics

141154